2017
The impact of disasters and crises on agriculture and food security

Food and Agriculture Organization of the United Nations
Rome, 2018

INTRODUCTION

Agriculture in an uncertain environment   1

PART I

More disasters, more impact on agriculture   13

©Reuters/Rodi Said

Heavily reliant on weather, climate, land, and water for its ability to thrive, agriculture is particularly vulnerable to natural disasters

One of the most direct ways in which natural disasters affect the sector is through reduced production. This results in direct economic loss to farmers, which can cascade along the entire value chain, affecting agricultural growth and rural livelihoods

The rising incidence of weather extremes will have increasingly negative impacts on agriculture because critical thresholds are already being exceeded

Disasters impact
agriculture beyond the short-term.
The sector often endures
long-lasting and multi-pronged
consequences such as loss of
harvest and livestock, outbreaks
of disease, and destruction
of rural infrastructure and
irrigation systems

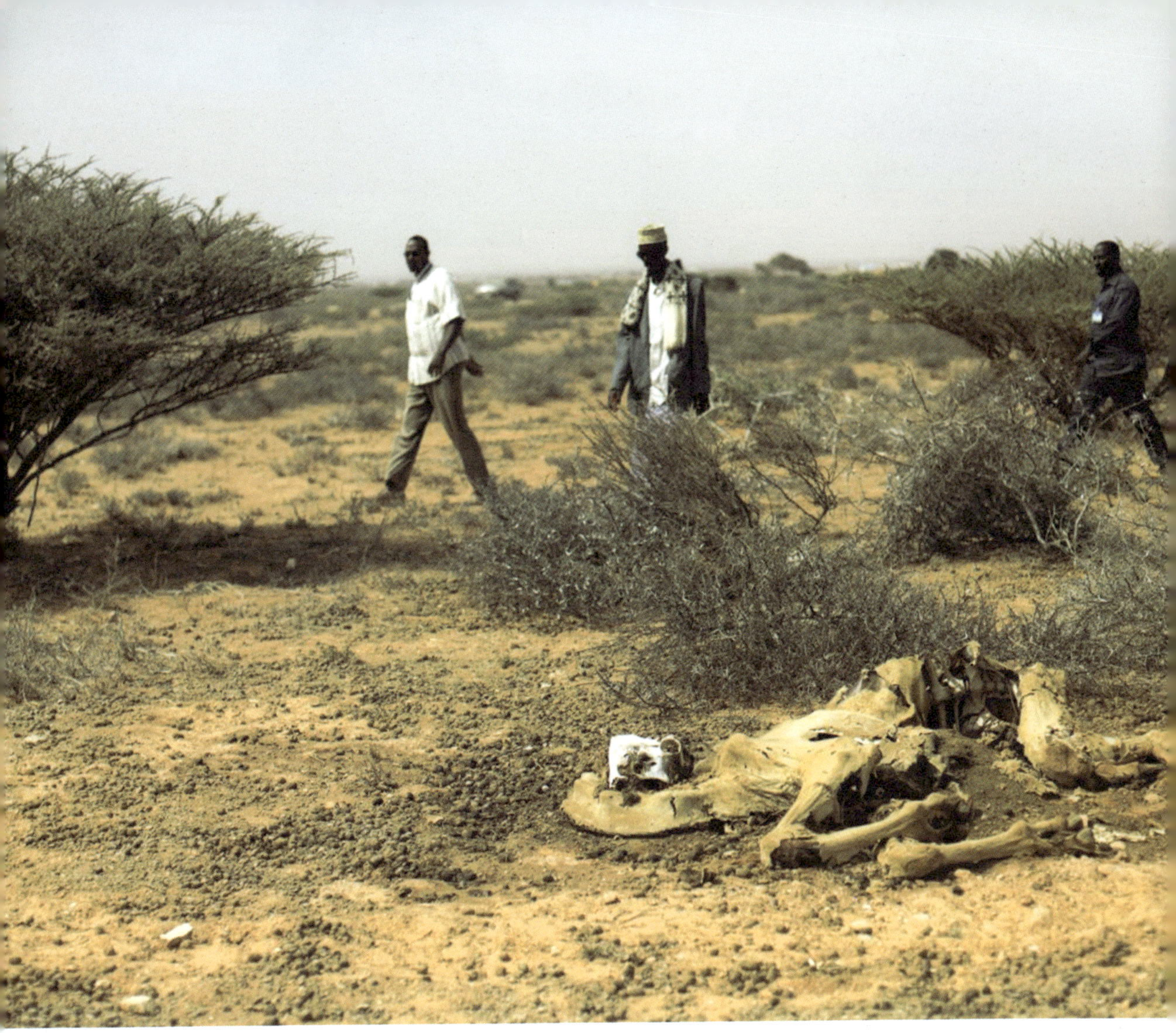

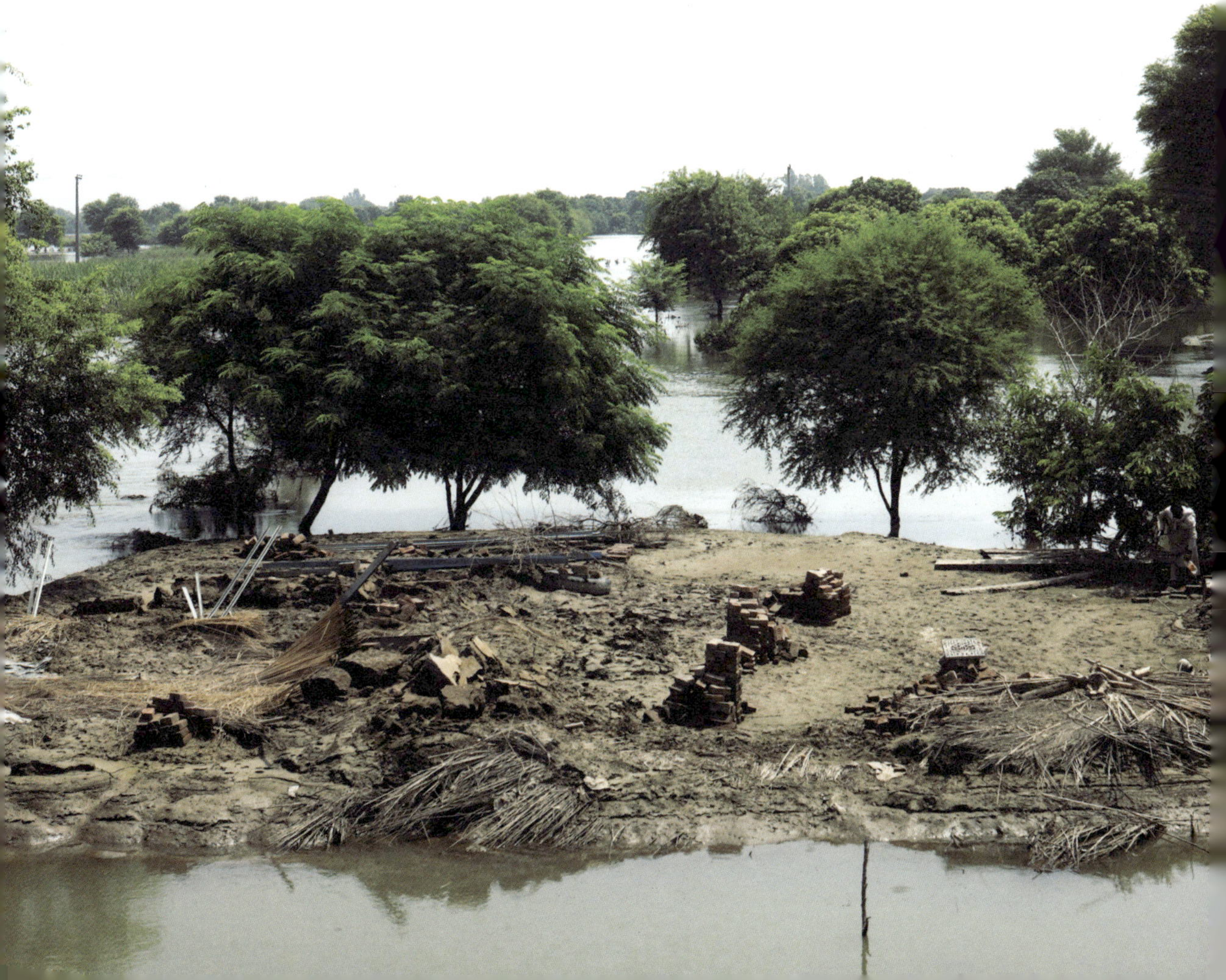

Natural disasters and protracted crises often overlap, aggravating impacts. More than 19 countries are currently affected by protracted crises, conflict and violence. Conflicts can devastate agriculture and rural livelihoods, causing significant economic loss, food insecurity and damage on all scales

Foreword

The second edition of FAO's report on the Impact of Disasters and Crises on Agriculture arrives at a pivotal moment for the future of disaster risk reduction and climate change. In 2017, the fifth Global Platform for Disaster Risk Reduction and COP23 brought the international community together to reiterate commitments for achieving targets of the SDGs, Paris Agreement and Sendai Framework for Disaster Risk Reduction. Global negotiations took place as the deadly trio of hurricanes Harvey, Irma and Maria wreaked havoc across developing and developed nations alike, a stark reminder that the 2030 Agenda for Sustainable Development cannot be achieved by countries while constantly battling flood waters and rebuilding flattened infrastructure.

Agriculture sectors face many risks, such as climate and market volatility, pests and diseases, extreme weather events, and an ever-increasing number of protracted crises and conflicts. Natural disasters have cost billions of dollars in lost agricultural production. The human food chain is under continuous threat from an alarming increase in the number of outbreaks of transboundary animal and plant pests and diseases. Conflict and protracted crises are forcing more and more people into conditions of poverty, food insecurity and displacement. This has become the "new normal," and the impact of climate change will further exacerbate these threats and challenges. Disaster risk reduction (DRR) and management must therefore become an integral part of modern agriculture.

The ability of governments, international bodies and other partners to operate and cooperate in fragile and disaster-prone contexts will become a defining feature for achieving resilience and sustainability. FAO is fully committed to the implementation of the 2030 Agenda and – beyond its duties as a custodian agency for monitoring global progress under 21 of the SDG indicators – has entered a key partnership with the UN Office for Disaster and Risk Reduction (UNISDR) on a resilience-related target. This entails FAO's contribution, support and leadership in monitoring a composite global indicator on economic loss in agriculture caused by disasters, corresponding to SDG target 1.5.2 and Sendai Framework indicator C-2.

As the 2030 Agenda sets our common vision for a transformed world and we progress towards global targets, it is crucial to understand and act on the messages of this report. Agriculture often bears a disproportionate share of disaster impacts, many of which are borne directly by smallholders. As resources become increasingly scarce, objective evidence is needed to effectively target our investments in resilience, preparedness and mitigation.

This report provides the latest data on the impact of disasters and crises on agriculture sectors, combined with sound analysis of remaining gaps and challenges. Its attention is not limited to natural disasters alone, but includes the first-ever analysis of the effect on agriculture of conflict and food chain crises. The 2017 report also considers how the entire sector is impacted: not only crops and livestock, but also forestry, fisheries and aquaculture.

Partnerships are needed to foster effective disaster risk management systems. Governments, international organizations, civil society and the private sector have the opportunity and obligation to work together in their commitment to a safer future for agriculture and rural livelihoods. A culture of disaster impact monitoring and assessment is an integral part of promoting effective DRR policy and action. Both national and local capacities must be strengthened to cope with increasing risks and recurring shocks. Building a more holistic and ambitious disaster-resilience framework for agriculture is crucial to ensuring sustainable development – which is a cornerstone for peace and the basis for adaptation to climate change.

José Graziano da Silva
Director-General
FAO

.

Acknowledgements

This second report on the impact of disasters and crises on agriculture and food security is the outcome of extensive cross-departmental collaboration within the Food and Agriculture Organization of the United Nations (FAO) to enhance the resilience of agriculture-based livelihoods to natural disasters.

The study forms a critical part of FAO's work under Strategic Objective 5: "Increase the resilience of livelihoods to threats and crises." Significant technical inputs and advice were provided by various divisions within the Organization from the departments of: Technical Cooperation, Economic and Social Development, Agriculture and Consumer Protection, Fisheries and Aquaculture, Forestry, and Climate, Biodiversity, Land and Water. In addition, FAO country offices provided invaluable support in gathering national-level data where available.

The study and report were coordinated and supervised by Stephan Baas, Piero Conforti and Shukri Ahmed. Galimira Markova was coordinating lead author.

Central to the development of the report were the technical papers prepared and technically reviewed by the following FAO colleagues: Niccolò Lombardi and Selvaraju Ramasamy (CBC) for Chapter I; Galimira Markova (ESS) for Chapter II with key inputs from Orsolya Mikecz; Niccolò Lombardi and Galimira Markova for Chapter III; Shawn McGuire and Amandine Poncin (AGPM) for Chapter IV; Florence Poulain (FIAP) and Robert Lee for Chapter V; Sheila Wertz-Kanounikoff, Jonas Cedergren, Peter Moore, Shiroma Sathyapala, Mats Nordberg and Arturo Gianvenuti (FOA) for Chapter VI; Ahmed Elldrissi (SP5), Juan Lubroth (AGAH) and Samuel Heft-Neal for Chapter VII; Neil Marsland, Daniele Barelli, Ulrich Nyamsi, Cara Kielwein and Brenda Lazarus (TCE) for Chapter VIII; Piero Conforti, Claude Raisaro and Galimira Markova for the Annex.

Editing, design and production supervised by Anne De Lannoy; editing by Laurie Olsen; design and layout by Claudia Neri, assistant Elisa Stagnoli.

Unless otherwise stated, all figures used in the publication are FAO.

Acronyms

ACAPS	Assessment Capabilities Project
CRED	Centre for Research on the Epidemiology of Disasters
COP	Conference of the Parties
DaLA	Damage and Loss Assessment methodology (ECLAC)
DL	Damage and Loss
DRM	Disaster Risk Management
DRR	Disaster Risk Reduction
EU	European Union
EM-DAT	Emergency Events Database (maintained by CRED)
ECLAC	Economic Commission for Latin America and the Caribbean
ESCWA	Economic and Social Commission for Western Asia
FAO	United Nations Food and Agriculture Organization
FEWS NET	Famine Early Warning System Network
FSNAU	Food Security and Nutrition Analysis Unit of FAO
GDP	Gross Domestic Product
GFDRR	Global Facility for Disaster Reduction and Recovery
ICIMOD	International Centre for Integrated Mountain Development (Nepal)
IFAD	International Fund for Agricultural Development
IPC	Integrated Phase Food Security Classification
IRRI	International Rice Research Institute
LTA	Long-Term Average
MEA	Millennium Ecosystem Assessment
OCHA	United Nations Office for the Coordination of Humanitarian Affairs
PDNA	Post-Disaster Needs Assessment
PPP	Purchasing Power Parity
SDG	Sustainable Development Goals
SFDRR	Sendai Framework for Disaster Risk Reduction 2015-2030
SIDS	Small Island Developing States
TAD	Transboundary Animal Disease
TLU	Tropical Livestock Unit
UNDESA	United Nations Department of Economic and Social Affairs
UNDG	United Nations Development Group
UNHCR	United Nations High Commissioner for Refugees
UNISDR	United Nations International Strategy for Disaster Risk Reduction
UNOCHA	United Nations Office for the Coordination of Humanitarian Affairs
WFP	World Food Programme
WHO	World Health Organization

Khulna, Bangladesh 2010 Women collecting drinking water from a communal water pump

Introduction
Agriculture in an uncertain environment

In 2015 FAO issued its first report on *The Impact of Disasters on Agriculture and Food Security*, exploring the negative effects of naturally-induced and climate-related disasters on agriculture. Against the backdrop of increasingly pressing challenges, FAO has now expanded the report's scope. The 2017 edition takes into consideration all threats facing agriculture today – from natural disasters in a changing climate to food chain crises and complex protracted crises, including civil conflict and war.

A vast number of agricultural livelihoods are compromised each year by disasters and crises. Smallholder family farms, which subsist on the production, marketing and consumption of crops, livestock, fish, forest products and other natural resources, need to cope in an environment of increasing volatility. Disasters can strike suddenly – such as earthquakes or a violent *coup d'état* – or develop slowly over time, as in the case of droughts. **Disasters can occur in isolation, in triggered consecutiveness or in simultaneous combination, with mutuallymagnifying effects. Such emergencies pose serious challenges to agricultural production and food security.**

Threats and crises

→	**NATURAL HAZARD-INDUCED DISASTERS** Henceforth referred to as natural disasters, those considered in this report are: geophysical (earthquakes, tsunamis and mass movements); droughts; floods; storms (including tropical, extra-tropical and convective); wildfires; extreme temperatures; biological disasters (epidemics, infestations).
→	**FOOD CHAIN CRISES** Transboundary plant, forest, animal, aquatic and zoonotic pests and diseases, food safety events, radiological and nuclear emergencies, dam failures, industrial pollution, oil spills, and so on.
→	**CONFLICTS AND PROTRACTED CRISES** -*conflicts*, e.g. civil unrest, regime change, interstate conflicts, civil wars -*protracted crises*, which develop as complex and prolonged emergencies and combine multiple types of conflict with other shocks, such as climate change.

The nature, frequency, intensity, and duration of a disaster determines its impacts on different entities, with smallholder farmers and the poor in both urban and rural areas disproportionately affected. Reinforcing the ability of such communities and institutions to prevent or mitigate the impacts of disasters – as well as to recover from and adapt to them in a timely, efficient and sustainable manner – is at the core of FAO's work on Disaster Risk Reduction (DRR). In 2015 FAO issued its first report on *The Impact of Disasters on Agriculture and Food Security*, exploring the negative effects of natural hazard-induced and climate-related disasters on agriculture. Against the backdrop of increasingly pressing challenges, FAO has now expanded the report's scope. The 2017 edition takes into consideration all threats facing agriculture today – from natural disasters in a changing climate, to food chain crises and complex protracted crises, including civil conflict and war.

In 2015, FAO issued its first report on The Impact of Disasters on Agriculture and Food Security, exploring the negative effects of natural hazard-induced and climate-related disasters on agriculture

Natural disasters and food chain crises

Since 1980, natural disasters have hit every continent and region of the world with growing frequency and intensity. The number of recorded natural disasters, along with their associated impact on livelihoods and economies at both local and national level is increasing significantly (Figure 1). These include geophysical disasters, climate and weather-related disasters as well as outbreaks of animal and plant pests and diseases (the biological disasters behind food chain emergencies). **On a global level, the economic loss associated with such disasters now averages between USD 250 billion to USD 300 billion every year.**[1] In developing countries, an average of 260 natural disasters occurred per year between 2005 and 2016, taking the lives of 54 000 people on average each year, affecting over 97 million others and costing an average of USD 27 billion in economic loss annually (EM-DAT CRED).

---

1   United Nations Office for Disaster Risk Reduction (UNISDR), 2015a, *Making Development Sustainable: The Future of Disaster Risk Management. Global Assessment Report on Disaster Risk Reduction.*

Figure 1. Occurrence of natural hazard-induced disasters in developing countries, 1980 – 2016

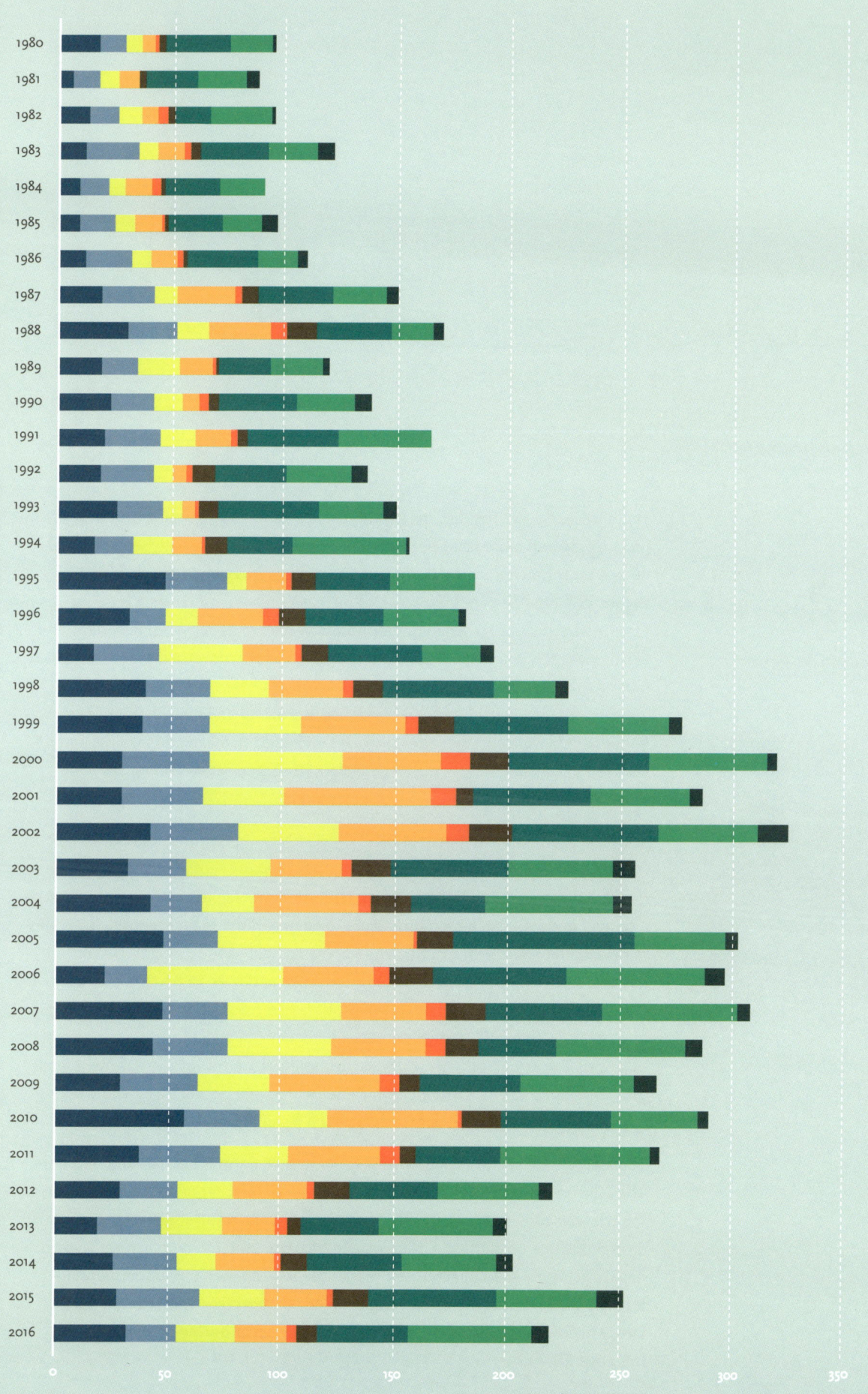

Legend: **Caribbean & Central America,** South America, East Africa, West and Central Africa, Southern Africa, Near East & North Africa, **South Asia, Southeast Asia, Oceania (Pacific SIDS)**

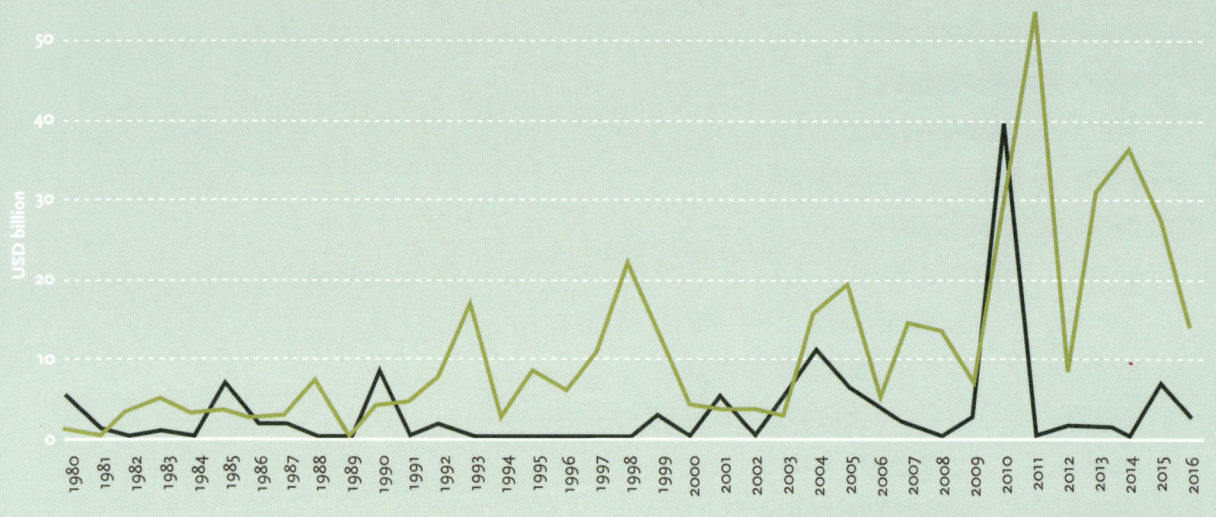

Figure 2. Economic loss from disasters in developing countries: geophysical vs. climate- and weather-related disasters, 1980–2016

Legend: **Geophysical Disasters, Climate-and Weather-Related Disasters**   *Source: EM-DAT CRED*

There were 260 natural disasters (both climate- and weather-related as well as geophysical and biological) per year in developing countries between 2005 and 2016 – an 11 percent increase on the 1993–2004 period when the average was 235 per year, and a more than two-fold increase on 1981–1992 when they averaged 122 per year (Figure 1). While the economic impact of geophysical disasters (earthquakes, tsunamis, volcanic eruptions and mass movements) has remained fairly stable over the past decades, annual economic loss from climate and weather-related events has been consistently growing, in line with the increasingly frequent occurrence of the latter (Figure 2). **Though damage and loss have not yet been calculated, 2017 – the most violent hurricane season on record – will certainly confirm this trend.**

The more immediate impacts of natural disasters – in terms of loss of human lives and destruction of critical infrastructure – occupy central space within the disaster discourse. **The impacts on agriculture, however, are seldom quantified or analysed in depth. Yet agriculture tends to be one of the main economic activities in developing countries,** contributing on average between 10 and 20 percent of national GDP in lower-middle-income countries and over 30 percent in low-income countries.[2] In some cases in Africa, this share can reach up to 39 percent (Niger) or 41 percent (Ethiopia, Mali).[3] Moreover, sustainable agriculture plays a key role in balancing the social, economic and environmental aspects of development while providing durable employment, sufficient income and decent living and working conditions for smallholder farmers and rural populations.

The agricultural sector is particularly vulnerable to natural hazards and disasters.[4] The observed notable increase in the frequency of weather-related events over the past decades poses a significant challenge to agricultural systems, given their crucial reliance on climate. Disasters can be detrimental to crop growth, livestock health, fisheries and aquaculture production and can seriously undermine the forestry sector. **Droughts cause long-term water shortages and extreme heat stress in crops, which can damage yields if they occur during certain times of the plant life-cycle.**

2   Country income groups as defined by The World Bank Atlas method.
3   World Bank, 2017, World Development Indicators.
4   Throughout this report, the term "agriculture" refers not only to crop and livestock but also to the subsectors of forestry, fisheries and aquaculture.

Combined with socio-economic factors and often with conflict, droughts have been responsible for some of the most serious famines in the world. Flooding can erode topsoil from prime growing areas, resulting in irreversible habitat damage. Storms, hurricanes and tornadoes can destroy forests and damage irrigation systems, silos, barns as well as other structures involved in agriculture production.

Furthermore, **the human food chain is under continuous threat from an alarming increase in the number of outbreaks of transboundary animal and plant pests and diseases.** Avian influenza (HPAI), peste des petits ruminants (PPR), locust infestations, plant pests, food-borne pathogens and mycotoxins are just some examples of food chain threats that have detrimental effects on food security, human health, livelihoods, national economies and global markets. PPR alone costs an estimated USD 1.45 billion to USD 2.1 billion each year. Plant diseases such as wheat rust can cause yield loss of up to 80 percent, putting worldwide wheat production at risk, while locust plagues have been known to destroy crops on a massive scale. This report takes into consideration food chain crises and explores the interplay between transboundary animal diseases, natural disasters and livestock production, while examining the case of Rift Valley Fever in Kenya (Chapter 7).

Transboundary animal diseases can cause production and economic loss estimated at USD 1.45 billion to USD 2.1 billion each year

Conflicts and protracted crises

Protracted crises are becoming the new norm, with 40 percent more ongoing food crises considered to be protracted than in 1990. Often driven by a combination of recurring causes, such as human-induced factors, natural hazards and lengthy food crises, protracted crises form a particularly challenging context in which to develop agricultural systems and fight hunger and malnutrition. As these crises persist, countries and communities need more effective and sustainable strategies to build their coping capacity against shocks and stressors. Nearly half a billion people live in 19 countries with protracted crises, mostly in Africa.

Impact of protracted crises

→ **Nearly half a billion people live in 19 countries** with protracted crises, mostly in Africa.

→ Of the 815 million people worldwide who suffer from chronic hunger, 146.6 million live in areas affected by protracted crises, 489 million live in conflict areas.

→ In 2016, the mean prevalence of undernourishment in protracted crisis situations was **30 percent compared with 10.8 percent on average in the rest of the developing world.**

→ **The majority of humanitarian assistance 2005–2015 was directed at protracted crises,** including those in the Sudan, Iraq, Afghanistan, Somalia as well as Palestine.

Resilience has emerged as a viable framework for integrating humanitarian and long-term development initiatives

Insufficient governance and institutional capacity to deal with the resulting challenges pose a serious threat to livelihoods and food systems. In a crisis context, undernourishment can be severe and levels of stunting and under-five mortality rates and particularly high. Resilience has emerged as a viable framework for integrating humanitarian and long-term development initiatives.

Countries in protracted crises require special attention, given the exceptional role that agriculture, natural resources and the rural economy play in people's survival as well as the damage to food and agriculture systems caused by such crises. This report represents a first-ever attempt at systematically quantifying the damage and loss on agriculture caused by protracted crises by exploring the case of the conflict in the Syrian Arab Republic (Chapter 8).

Given the increasing frequency and gravity of the main categories of threats and the risks they pose to agriculture, it is crucial to develop adequate disaster and crisis governance structures – including enabling policies, strengthened capacities and targeted financing mechanisms to counteract the impacts. To this end, bridging the knowledge gap is the first step. Estimating and quantifying the impact of natural disasters, climate-related events, food chain hazards and protracted crises on the agricultural sector is crucial to protecting the investments made in development and strengthening resilience. Too often however, methodical documentation of the economic impact of disasters on agriculture is either non-existent or insufficient, and/or the documentation that does exist is under-analysed.

More disasters, but dearth of information on damage and loss in agriculture

Notwithstanding the many efforts at both national and international levels, there is limited information on the impact of disasters, food chain crises and conflict on agriculture and its subsectors – crop, livestock, fisheries, aquaculture and forestry. This is largely because data on the agricultural impact of disasters is not collected or recorded in a systematic way, i.e. by subsector and at regional, national and subnational level. Globally available statistics on damage and loss from disasters do not offer a sufficient level of disaggregation in order to allow for an in-depth understanding of the mechanisms at play.

Globally available statistics on damage and loss from disasters do not offer a sufficient level of disaggregation in order to allow for an in-depth understanding of the mechanisms at play

Post-disaster needs assessments (PDNAs), typically conducted in the aftermath of large-scale disasters to inform humanitarian responses, do provide some evaluation, mostly of immediate effects across relevant sectors.[5] However, needs assessments do not share a common method for estimating damage and loss (some use livelihood or food economy approaches, while others focus on the economic impact of physical damage), thus creating different perspectives on agricultural impact. More often than not, the resulting data is not systematically included in national disaster databases. **Both the long-term consequences and the evolution of the disaster's impact on the sector are poorly understood at the national, regional and global levels.**

This calls for the establishment of a more robust evidence base. A thorough analysis of existing trends in agricultural production and related impacts on production volumes and patterns due to disasters is a key starting point. Such analysis can bridge the information gap and inform decision-making for DRR, sustainable development and emergency response planning. A robust, sector-specific damage and loss data inventory is instrumental for designing effective DRR policy and practice. National strategies for DRR and climate change adaptation that support resilience and sustainable agricultural development must be informed by the particular nature of the disaster's impact on the sector.

National strategies for disaster risk reduction and climate change adaptation that support resilience and sustainable agricultural development must be informed by the particular nature of the disaster's impact on the sector

The cumulative impacts of natural disasters, climate-related events, food chain hazards and protracted crises will ultimately depend on adaptive, smart national and international strategies and polices, on shifting global market conditions as well as on local responses to climate stressors. The threats that pose the greatest risk and the most severe loss (such as disasters related to climate change, conflicts, etc.) must be systematically addressed at all levels in order to effectively counteract the destabilizing impacts on sector growth and food security.

---

5 A PDNA is a system of processes and methods used to assess, plan and mobilize support for the recovery of countries and populations affected by disasters. Typically, the process is owned and led by the government and supported by UN Agencies, the EU and The World Bank.

Ultimately, this will support government efforts to achieve sustainable agricultural development, alleviate hunger and move closer to meeting the pledged international targets, including the Sendai Framework and the SDGs.

The Sendai Framework recognizes the crucial role of local governments, UN agencies, international and national organizations in reducing disaster risk

Adopted in March 2015, the Sendai Framework recognizes the importance of reducing disaster risk and the crucial role of local governments, UN agencies, international and national organizations in tackling the challenges. The Framework further links DRR to broader resilience targets, such as the SDGs. The 2030 Agenda for Sustainable Development stipulates that all countries, regardless of their income level, should be prepared to effectively prevent and mitigate any disaster impacts. Where disasters cannot be avoided, efforts should be made to minimize their devastating effects on livelihoods and the economy. DRR strategies are essential to ensure that increasingly frequent hazardous events do not push countries and communities into poverty loops. Political and financial backing for DRR must be mobilized through adequate policy frameworks. The role **of agriculture in these efforts is fundamental, given its wide interactions with the environment, its direct reliance on natural resources for production and its crucial role in national socio-economic development.**

As documented throughout this report, FAO's dedicated focus on improving assessment of the impacts of climate-induced extreme events on agriculture also serves to inform and enrich the future climate change adaptation agenda, namely by supporting the Paris Agreement indicators and further advancing the goals of the Warsaw International Mechanism for Loss and Damage associated with Climate Change Impacts.

Purpose and scope of this report

The Impact of Disasters on Agriculture and Food Security 2015 showed that a staggering 22 percent of total damage and loss from natural disasters in developing countries was absorbed by the agriculture sector alone.

This FAO report updates the state of post-disaster agriculture in developing countries. It presents a first-ever, in-depth analysis of disaster impact on the subsectors of fisheries, aquaculture and forestry

Two years on, FAO continues the effort to bridge persisting knowledge gaps and foster a better understanding of how the agriculture sector is affected by disasters. Through this 2017 report, FAO refreshes its 2015 conclusions and provides an update on the state of post-disaster agriculture in developing countries. It presents a first-ever, in-depth analysis of disaster impact on the subsectors of fisheries, aquaculture and forestry, which are not always covered by PDNAs; It also reveals an agriculture-specific methodology for evaluating damage and loss from disasters, thereby improving understanding of the wider implications for the economy and livelihoods. **Finally, the report also looks at all threats facing agriculture, including food chain crises and transboundary animal diseases – which are increasingly common and tend to have multipronged impacts on agriculture – as well as conflict and protracted crises, which are also on the rise.** The latter is accomplished through an analysis of the impact on the agricultural system and rural livelihoods in the Syrian Arab Republic.

Furthermore, FAO seeks to continue providing updated and systematic data and analysis in order to build a holistic information system on the impact of disasters and crises on agriculture in developing countries. By systematically improving disaster damage and loss assessment, FAO's work will directly contribute to implementing and monitoring the two main 2015 international agendas, which recognize resilience as fundamental to their achievement, namely the SDGs and the Sendai Framework. The newly developed methodology for assessing damage and loss from disasters in agriculture, introduced in the Annex, aims to improve agriculture-related resilience monitoring within the UN-wide system by providing a standardized set of procedural and methodological steps that can be used at global, national and subnational levels. This will enable thorough damage and loss assessment in the sector, ensuring consistency across countries and disasters. The new FAO methodology has already been adopted by UNISDR to help monitor the achievement of specific targets in the Sendai and SDG frameworks for reducing economic loss from disasters.[6]

FAO's new methodology will be used by UNISDR to monitor progress toward specific Sendai Framework and SDG targets

While the importance of their impacts is undisputed, **natural disasters continue to pose various methodological conundrums, such as the debate on how to define a disaster and classify it accordingly.** The classification used throughout this report is in line with established Centre for Research on the Epidemiology of Disasters (CRED) and UNISDR classifications,[7] which include the main disaster sub-groups of geophysical, meteorological, hydrological, climatological and biological disasters. Special emphasis is placed on the disaster types in each sub-group that are of particular significance to agriculture, for example droughts, floods, extreme temperature, storms, diseases, and so on.

Furthermore, the concepts of hazards, risks, vulnerabilities and disasters must be framed from an agricultural perspective, while relying on universal definitions. **This report retains established UNISDR terminology on DRR for defining the main concepts related to disasters.** While some have been adjusted for use in the agricultural context, they remain compatible with UNISDR definitions.

Key terms

→ **Hazard:** a process or phenomenon that may cause loss of life, injury or other health impacts, property damage, social and economic disruption or environmental degradation. While hazards may be natural, anthropogenic or socio-natural in origin, this report refers to hazards of natural origin only.

→ **Hazardous event:** the occurrence of a natural phenomenon in a particular place during a particular period of time due to the existence of a hazard.

→ **Disaster:** a serious disruption of the functioning of a community or a society due to hazardous events interacting with conditions of exposure, vulnerability and capacity, leading to one or more of the following: human, material, economic and environmental loss and impacts.

→ **Disaster risk reduction (DRR):** the policy objective aimed at preventing new and reducing existing disaster risk and managing residual risk, all of which contributes to strengthening resilience.

→ **Damage:** the total or partial destruction of physical assets and infrastructure in disaster-affected areas, expressed as replacement and/or repair costs. In the agriculture sector, damage is considered in relation to standing crops, farm machinery, irrigation systems, livestock shelters, fishing vessels, pens and ponds, etc.

→ **Loss:** refers to the changes in economic flows occurring as a result of a disaster. In agriculture, loss may include decline in crop production, decline in income from livestock products, increased input prices, reduced overall agricultural revenues, higher operational costs and increased unexpected expenditures to meet immediate needs in the aftermath of a disaster.

---

6   The FAO methodology will be used to monitor progress towards achieving SDG target 1.5 on building resilience and reducing vulnerability to climate-related extreme events and disasters (in particular by measuring agriculture-related components of indicator 1.5.2 on reducing direct disaster economic loss). Similarly, the methodology will be used to measure SFDRR indicator C-2 to reduce direct agricultural loss attributed to disasters. This is done through a collaborative process between FAO and UNISDR, the custodian UN agency for the above-mentioned targets.

7   The classification of disasters adopted here draws on the EM-DAT CRED Guidelines and is also in line with UNISDR definitions and terminology.

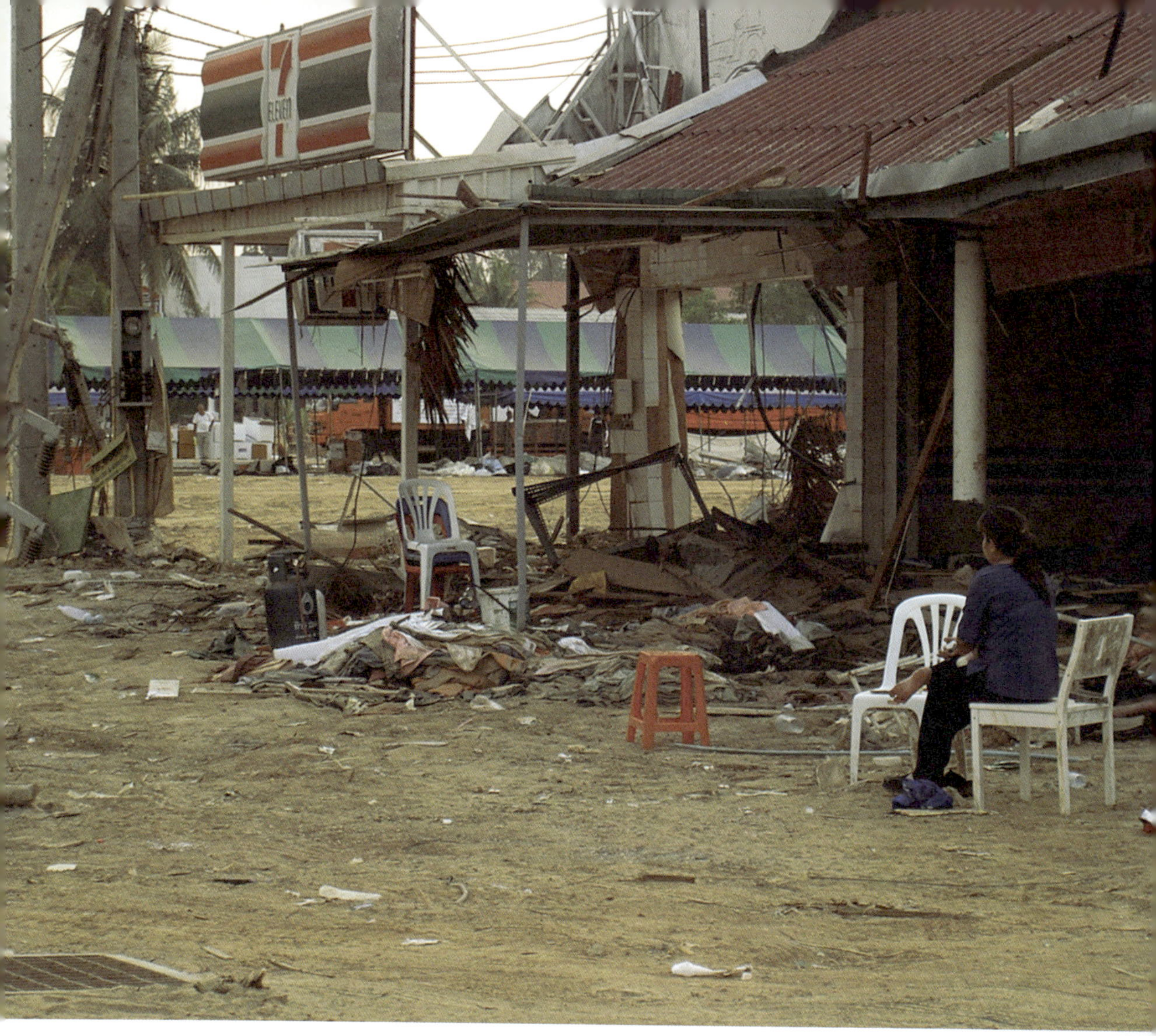
7 ELEVEN

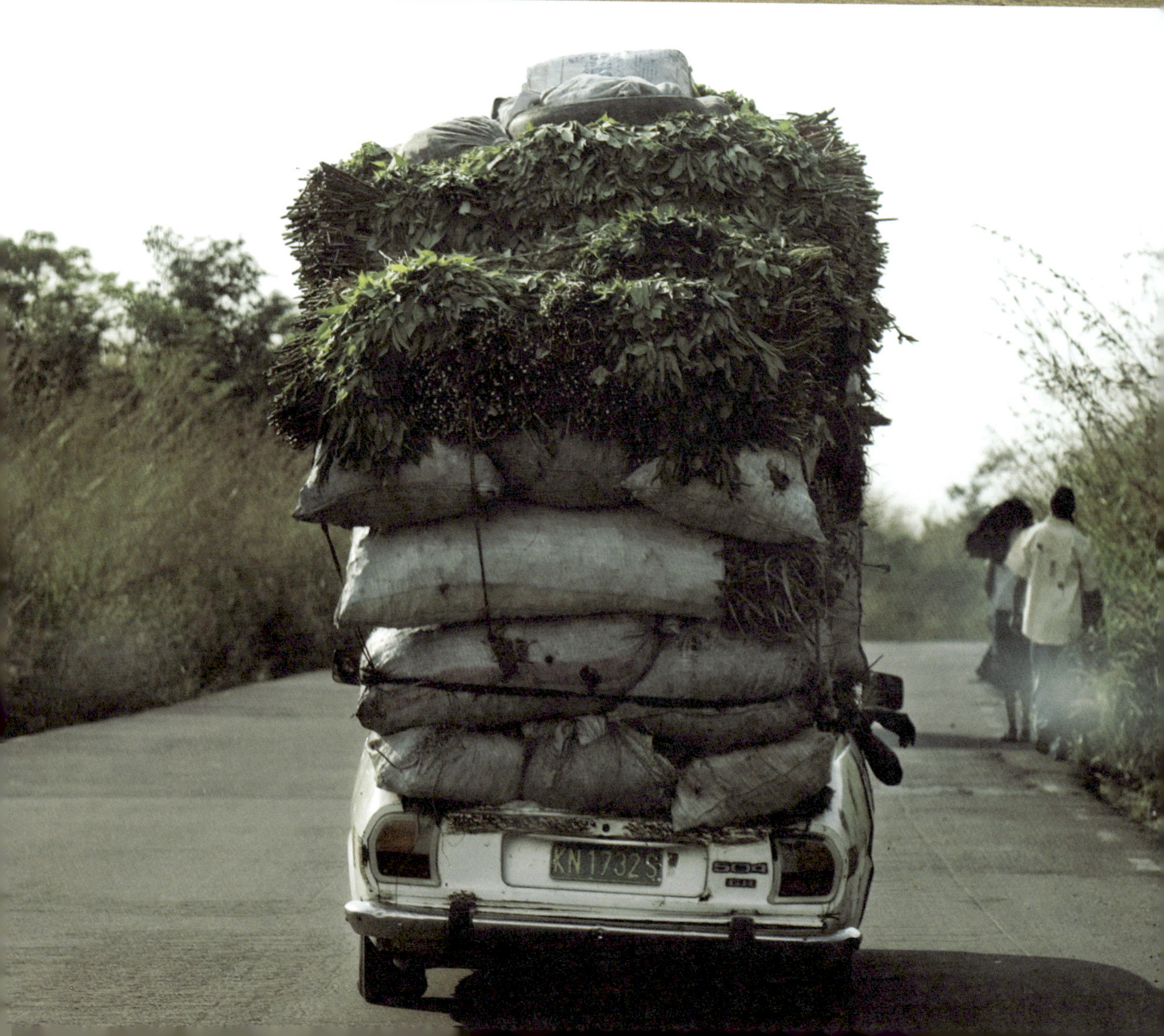
KN1732S

The report comprises three sections:

→ **Part I.** More disasters, more impact on agriculture
The first part of the report explores the breadth and scope of the impact of natural disasters on the agriculture sector. Through comparative analysis of PDNA findings, Chapter 1 places agriculture on the map of post-disaster economic disruption and identifies its relative share of overall impact. Furthermore, it identifies how damage and loss is distributed across subsectors and according to disaster type. Chapter 2 takes a bird's eye view of how agricultural production is affected by natural disasters, examining the extent of crop and livestock production loss in developing countries over the past decade. It takes both a global and regional perspective and presents an improved analysis, taking into account the effect of both large- and medium-scale disasters as well as smaller-scale events. Overall, the results presented in Part I provide a grounded understanding of the economic consequences of natural disasters for agriculture in order to inform adequate DRR policy and action.

→ **Part II.** Estimating damage and loss: getting it right
Shifting the focus from cumulative natural disaster impact to measuring the direct effects of individual events, Part II presents the state of affairs on damage and loss assessment in agriculture. FAO's standardized methodology aims to form the backbone of disaster impact assessment in all agricultural subsectors. Chapter 3 lays its foundations and tests its application through a case study-based analysis of two diverging types of disasters – Typhoon Haiyan in the Philippines and past drought occurrences in Ethiopia. Chapter 4 takes the analysis further. It explores the importance of household-level data for adequate damage and loss assessment, using the Nepal earthquake of 2015 as a case at hand. Building local capacity for data collection is key to the successful application of FAO's methodology. On the other hand, gaps in disaster-related data at either household or macro- level can significantly hinder impact assessment. This is particularly relevant in the case of fisheries and aquaculture (Chapter 5) and forestry (Chapter 6), the two subsectors that often remain on the fringes of damage and loss assessment, threatening to grossly undermine understanding of the impact borne by them. The report investigates the structures necessary for an effective damage and loss assessment in fisheries and forestry that informs adequate DRR policy and action.

→ **Part III.** Covering new ground: food chain crises, protracted crises, conflict
This section extends the analysis beyond natural disasters and provides a first glimpse of the effects that other types of threats have on agriculture. Chapter 7 explores the growing frequency and severity of Transboundary Animal Disease outbreaks and their implications for the livestock subsector and for the human food chain. The foundations are laid for an integrated analysis of damage and loss caused by animal and zoonotic diseases on the livestock sector, which is imperative for the implementation of effective policies and action to prevent or limit the geographic spread of animal diseases, minimize their impact and respond to food chain emergencies. Chapter 8 is devoted to protracted crises. With a growing incidence and prolonged duration, they form a particularly challenging context for people, their agricultural systems and food security. In the Syrian Arab Republic, FAO has pioneered an adapted approach to assess agricultural damage and loss in the context of conflict. This chapter offers a first insight into using crisis impact assessment to inform reconstruction and humanitarian response in agriculture.

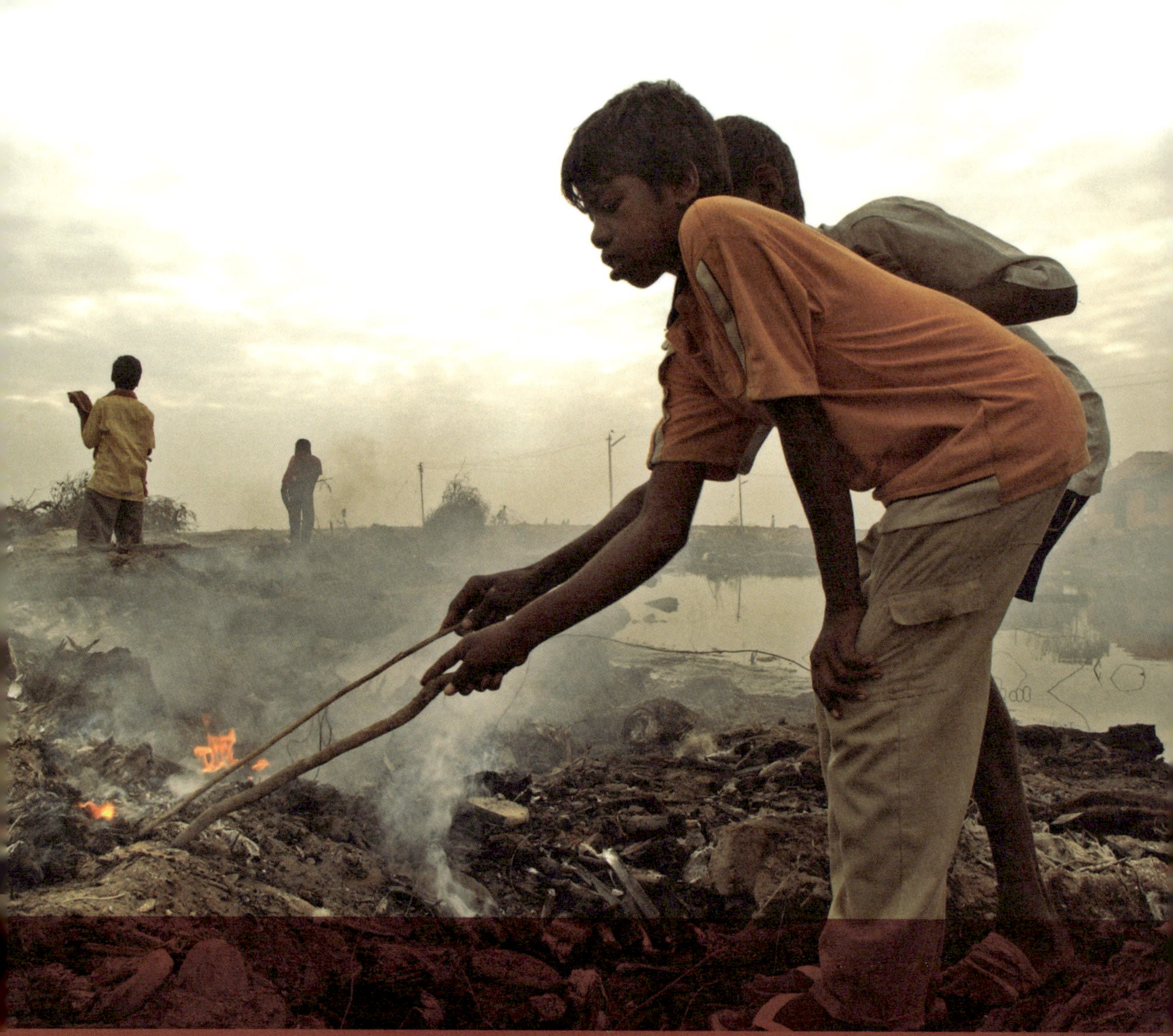

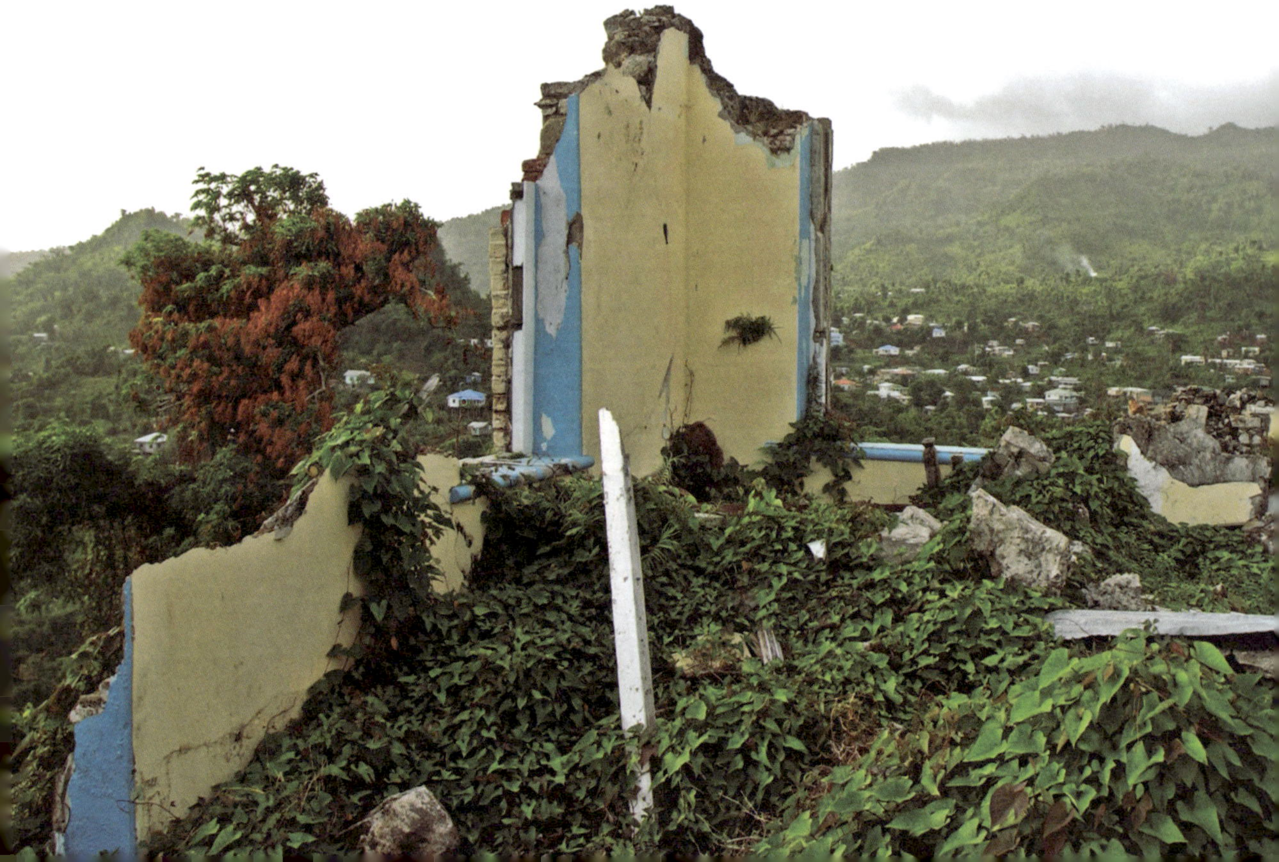

PART I ⟨ More disasters, more impact on agriculture

Iraq 2017 ⟨ Displaced civilians flee fighting between Iraqi forces and ISIS with their livestock

Chapter I
Disaster damage and loss – a hefty share for agriculture and its subsectors

This chapter places agriculture on the map of post-disaster economic disruption. Based on PDNA findings, the relative share of damage and loss in agriculture over the past decade is derived as well as the particular sub-sectoral impacts on crops, livestock, forestry, fisheries and aquaculture. A great share of the overall brunt of disasters falls on agriculture, and each subsector is affected differently by different types of hazards. Beyond physical damage and economic loss, disasters often have far reaching effects on food security, natural resources and the ecosystem.

Relative damage and loss from disasters – where does agriculture stand?

A review of 74 PDNAs conducted in 53 developing countries over the past decade (2006–2016) shows that agriculture (crops, livestock, fisheries, aquaculture, and forestry) absorbed 23 percent of all damage and loss caused by medium- to large-scale natural disasters (Figure 1).[1] **When only climate-related disasters (floods, drought, tropical storms) are considered, the share of damage and loss absorbed by agriculture increases to 26 percent.**

Damage to agricultural assets accounts for 16 percent of damage in all sectors.[2] The destruction of facilities, machinery, tools, and key infrastructure related to agricultural production has a significant impact, especially on the most vulnerable, who may need a long time before being able to rebuild damaged assets and resume their productive activities.

Almost one-third of all disaster loss is accrued in the agricultural sectors. In countries where a large number of smallholders rely on agricultural production for their subsistence and livelihoods, declines in production flows pose serious threats to food security. These findings show that vulnerable farmers, herders and fishermen bear the brunt of disaster impacts.

83 percent of all damage and loss caused by drought was absorbed by agriculture

→ Drought affects the agriculture sector disproportionately, relative to other sectors (Figure 2): **83 percent of all damage and loss caused by drought was absorbed by agriculture.** The crop and livestock sectors are most affected by this slow-onset hazard.

→ Volcanic eruptions, storms, floods, tsunamis and earthquakes also have a major impact on the sector. Although earthquakes cause a relatively low impact on agriculture in general, they have severe negative consequences for rural livelihoods due to the high costs of rebuilding destroyed buildings and infrastructure. Nepal's 2015 earthquake, for instance, caused significant damage and loss to agriculture and increased the vulnerability of affected communities, especially women, to hunger and food insecurity (Government of Nepal, 2015a). This report takes a closer look at the agricultural impact of that earthquake in Chapter 4.

Damage and loss per sector

Between 2006 and 2016, crops, livestock, fisheries, aquaculture, and forestry absorbed 23 percent of all damage and loss

A comparative analysis of PDNAs across sectors shows that just under half of the impact of disasters on agriculture is absorbed by the crops sector, while 36 percent is absorbed by livestock (Figure 3).[3] The share of fisheries, aquaculture and forestry accounts for 7 percent, however these sectors often remain underreported in PDNAs. The impact of disasters on forestry is generally acknowledged in the assessments, although rarely quantified in monetary terms. Furthermore, over one-quarter of all disasters assessed through PDNAs occurred in Small Island Developing States (SIDS), where damage and loss in fisheries, albeit low in absolute terms, can have far-reaching consequences on the livelihoods of local fishing communities. The current report and the supporting FAO methodology for damage and loss assessment aim to address the prevailing information gaps and take a first-ever sector-specific approach. Chapters 5 and 6 present a brief overview of how the fisheries, aquaculture and forestry sectors are impacted by disasters and discuss implications for the sub-sectoral application of damage and loss assessment methodology.

---

1   This figure is consistent with the one calculated for the period of 2003–2013, when the sector absorbed 22 percent of total damage and loss (FAO, 2015b).
2   For a more detailed definition of the concepts of damage and loss, see the Introduction.
3   The "unspecified" category refers to damage and loss to agriculture for which the PDNAs do not provide a disaggregated figure by subsector.

Figure 1. Damage and loss in agriculture as share of total damage and loss in all sectors (2006–2016)

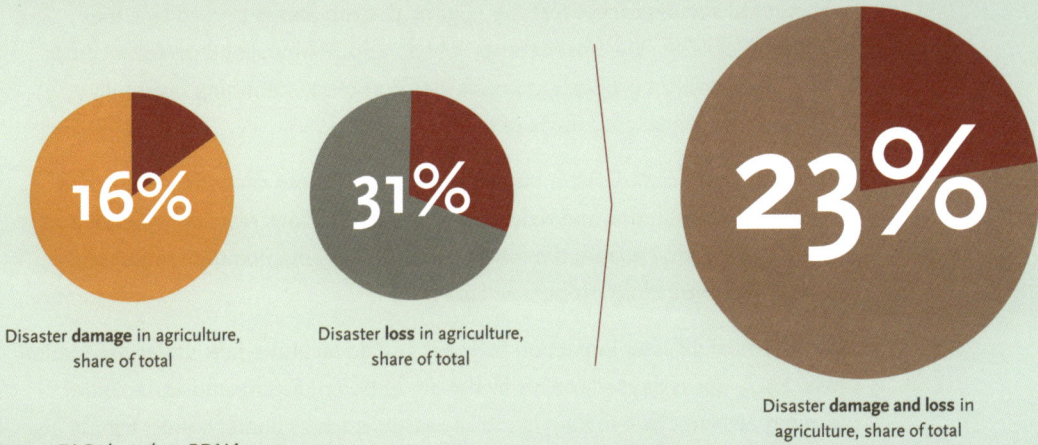

16%

Disaster **damage** in agriculture, share of total

31%

Disaster **loss** in agriculture, share of total

23%

Disaster **damage and loss** in agriculture, share of total

Source: FAO, based on PDNAs

Figure 2. Damage and loss in agriculture as share of total damage and loss across all sectors (2006–2016), by type of hazard

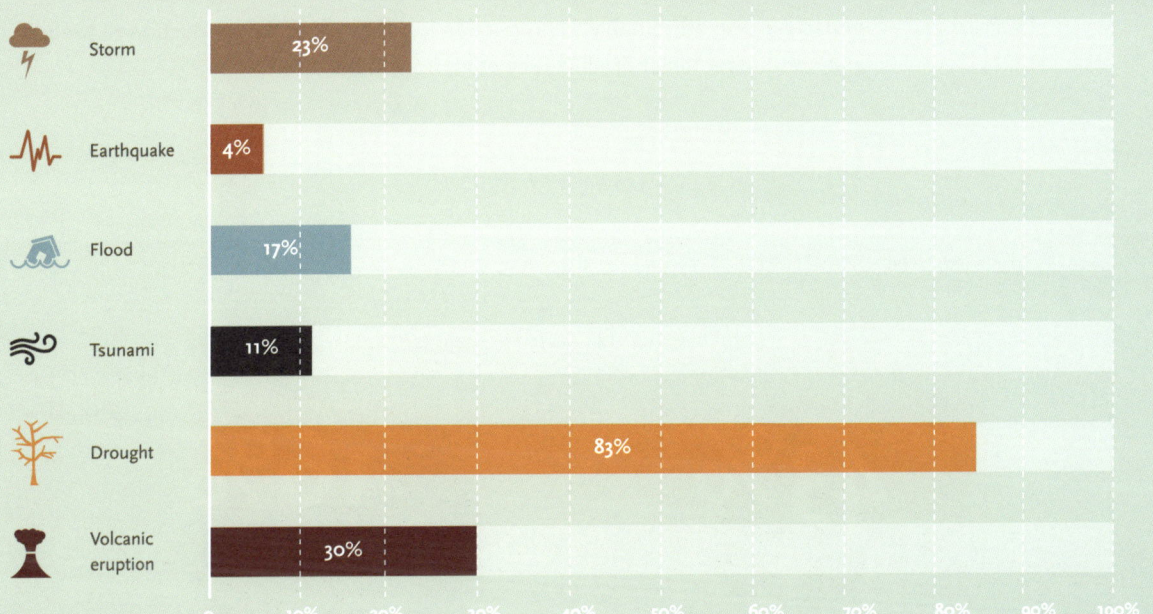

Hazard	Share
Storm	23%
Earthquake	4%
Flood	17%
Tsunami	11%
Drought	83%
Volcanic eruption	30%

Source: FAO, based on PDNAs

Figure 3. Damage and loss in agriculture by agricultural sub-sector, percentage share of total (2006–2016)

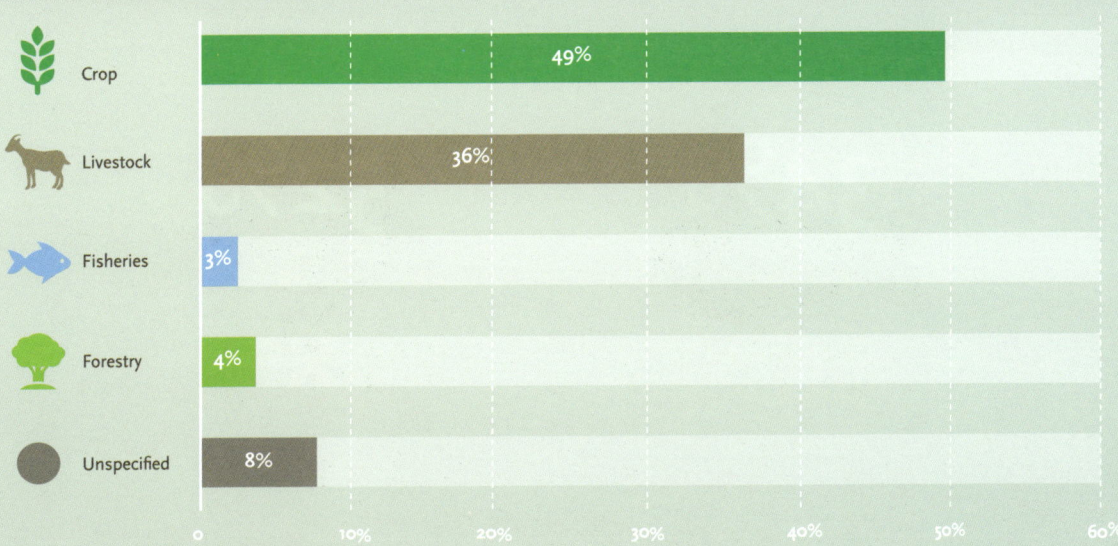

Sub-sector	Share
Crop	49%
Livestock	36%
Fisheries	3%
Forestry	4%
Unspecified	8%

Source: FAO, based on PDNAs

Damage versus loss

Production loss is greater than damage to assets and infrastructure in all agricultural sectors except forestry (Figure 4). Crop loss is caused by either sudden shocks or slow-onset events, which reduce annual and perennial crop yields. Longer-term production loss is also common, stemming from fully destroyed perennial crop fields (e.g. fruit trees).

The most significant disaster impact on livestock is loss caused by weakened animal body conditions and reduced animal productivity. Also, considerable livestock loss derives from the foregone value of production until re-stocked livestock becomes fully productive again.

Two-thirds of disaster impact on fisheries and aquaculture falls under production loss. Such loss is caused mostly by the disruption of fishing and aquaculture farming activities due to damage to key assets such as boats, ponds, fishing gear and hatchery farms, among others.

Finally, most of the impact of disasters on forestry is due to damaged forest trees broken and knocked down by violent tropical storms. A significant share of impacts is also attributable to production loss from declined production of timber and non-timber forest products resulting from shocks.

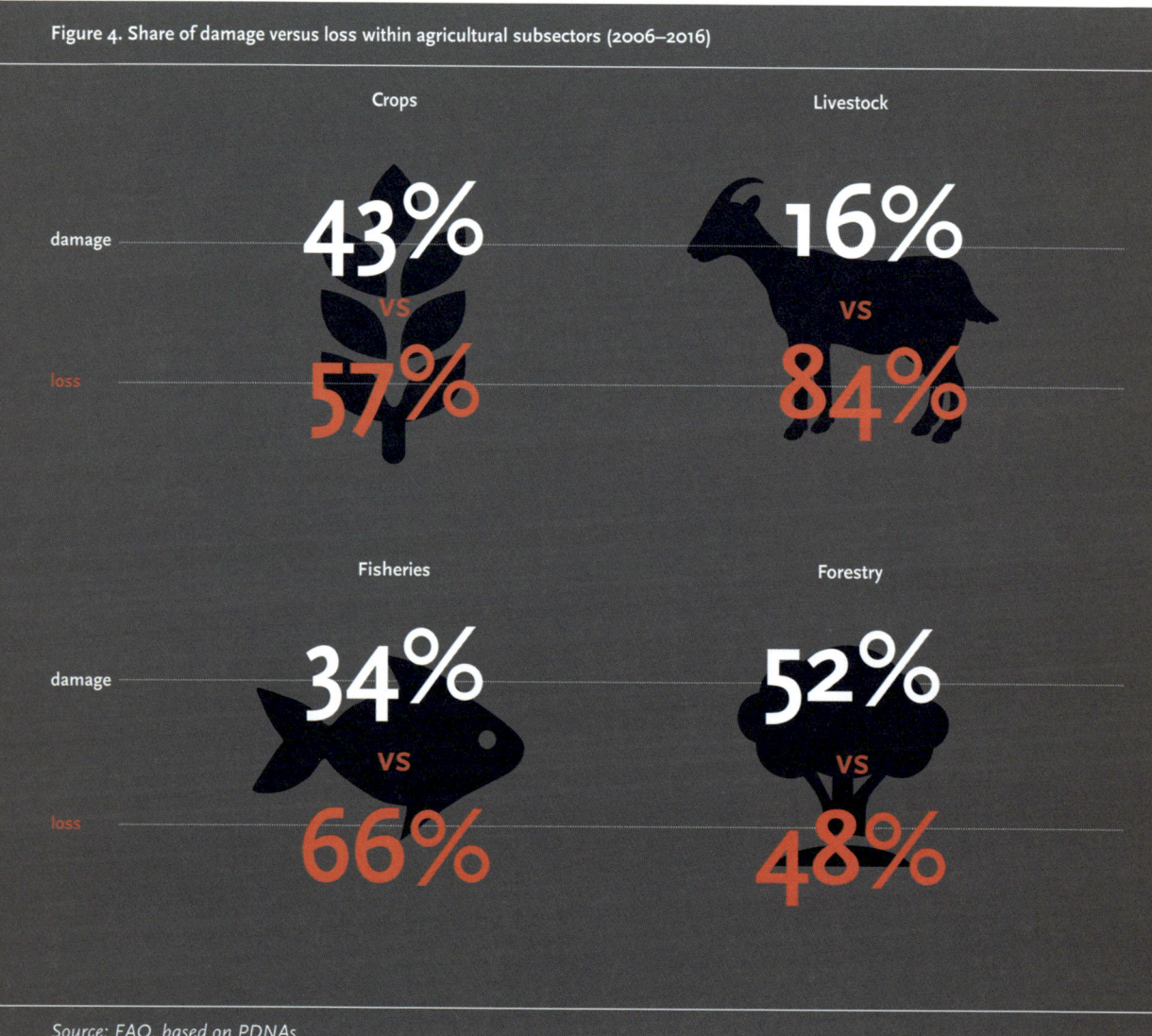

Figure 4. Share of damage versus loss within agricultural subsectors (2006–2016)

Crops

damage **43%**

vs

loss **57%**

Livestock

damage **16%**

vs

loss **84%**

Fisheries

damage **34%**

vs

loss **66%**

Forestry

damage **52%**

vs

loss **48%**

Source: FAO, based on PDNAs

Different disasters – different sector impacts

Agricultural sectors are affected differently by different types of hazards (Figure 5). From 2006 to 2016, almost two-thirds of all damage and loss to crops was caused by floods. In absolute terms, the most harmful disaster for crops was the 2010 flood in Pakistan (USD 4.5 billion), followed by the 2008–2011 drought in Kenya (USD 1.5 billion). In recent years, global crop production was severely affected by events such as the 2015 floods in Myanmar and the 2014 floods in Bosnia and Herzegovina. For Myanmar, damage and loss was USD 572 million, while for Bosnia and Herzegovina it was USD 255 million. In both cases, the cost occurred as a result of reduced yields and late planting due to limited access to arable land.

Drought remains by far the most harmful disaster for livestock, causing 86 percent of total damage and loss in the sector. The largest impact over the past decade is attributed to the 2008–2011 drought in Kenya (USD 8.9 billion) and in the overall Horn of Africa region.

Yet this is neither a recent nor an isolated phenomenon. The Horn of Africa has been synonymous with drought since the 1980s and it made headlines once again in 2017. The region is currently experiencing severe drought, which has triggered a humanitarian crisis of skyrocketing food insecurity, disease outbreaks and displacement, particularly among pastoral and agro-pastoral communities. Despite a targeted emergency response, humanitarian needs continue to rise, with 15 million people in need of emergency food assistance.

Figure 5. Damage and loss to agriculture sectors by type of hazard (2006–2016)

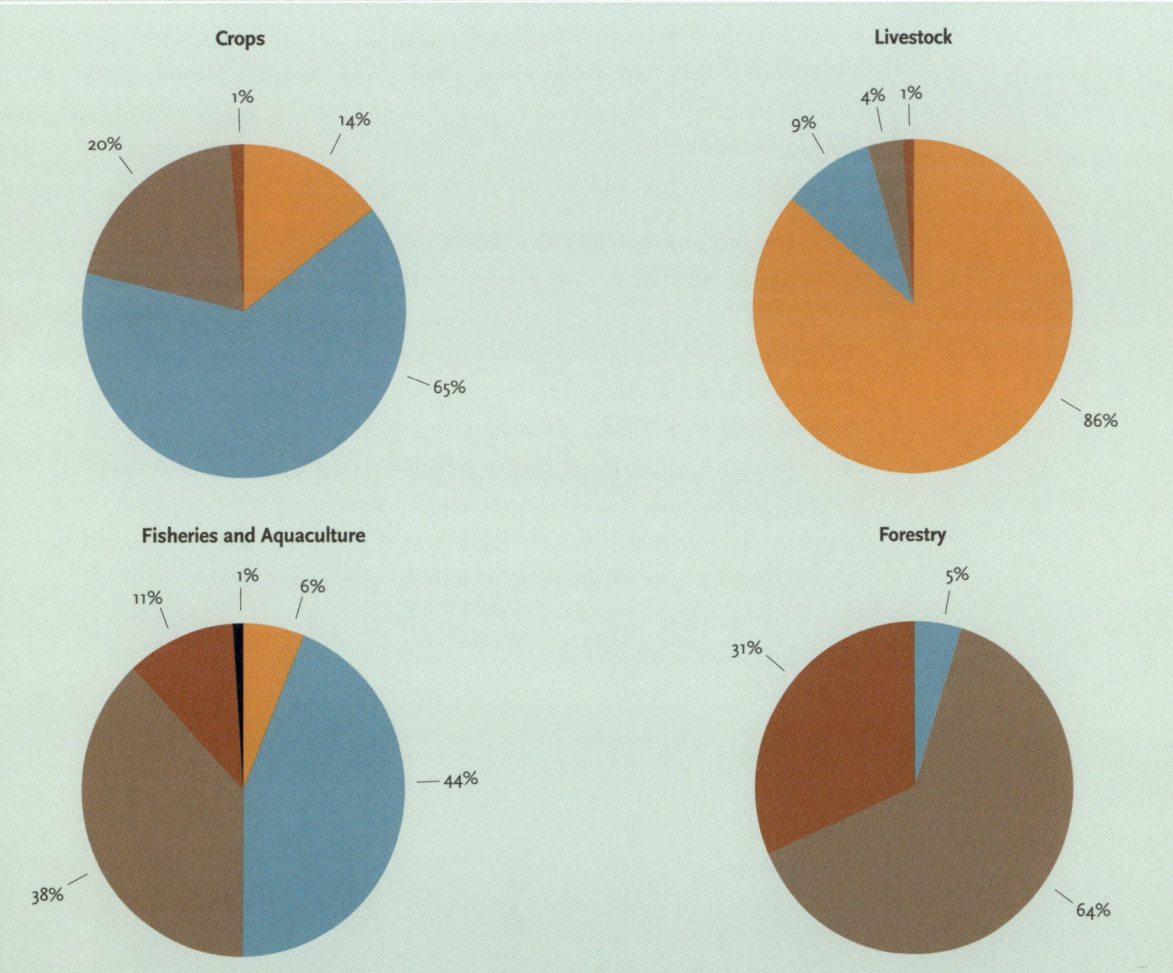

Legend: **Storms, Floods, Drought, Tsunamis, Earthquakes.**

Source: FAO, based on PDNAs

The fisheries and aquaculture sector is mostly impacted by floods and storms. The largest recorded impact over the past decade was caused by the Myanmar floods in July and August 2015, amounting to USD 240 million in subsector damage and loss. This was largely attributed to declines in aquaculture and inland fisheries production, although damage to fishing gear, equipment, boats, hatchery farms and aquaculture facilities was also significant.

Storms caused about two-thirds of all disaster impacts on forestry recorded in PDNAs between 2006 and 2016

Storms caused about two-thirds of all disaster impacts on forestry recorded in PDNAs between 2006 and 2016, especially due to the impact of Nicaragua's 2007 Hurricane Felix and Myanmar's 2008 Cyclone Nargis. In recent years, the 2015 earthquake in Nepal had a major impact on forestry, as it destroyed large areas of natural forests and caused a drop in the collection of non-timber forest products, affecting a large number of Community Forestry User Group (CFUG) members. **Total damage and loss caused by the Nepal earthquake on forestry amounted to about USD 308 million, corresponding to about 30 percent of all damage and loss recorded in the forestry sector in the analysed PDNAs.**

Impact of disasters on food security

Disasters pose direct and indirect threats to the livelihoods and food security of smallholder farmers. The number of people in need of food assistance often increases after the occurrence of disasters, especially when vulnerable populations are affected. The 2015–2016 El Niño-related droughts and floods, for instance, heavily affected the food security and nutritional status of more than 60 million people globally (FAO, 2016). The uncertainty associated with the observed increase in the frequency and intensity of disasters in many developing countries can drive poor farmers to invest in low-risk but low-returning agricultural production technologies and techniques (Cole *et al.*, 2013). In turn, low investments can lead to lower future farm profits and increased food insecurity (Aimin, 2010). Furthermore, the impact of disasters on reduced food consumption, education and healthcare can lead to long-term negative effects in terms of income generation and future food security (FAO, 2015c).

Food insecurity and disaster risk reinforce one another. Disasters have shattering consequences on food security, and food-insecurity increases vulnerability, leading to a downward spiral in which rural livelihoods are increasingly eroded (Garschagen *et al.*, 2015). In disaster situations, food-insecure people might find themselves forced to take desperate measures to address immediate needs, often compromising their livelihoods and increasing their vulnerability and exposure. **Severe droughts, for instance, can force food-insecure farmers to overexploit common property resources** such as community forests, pasture, ponds, riverbanks and groundwater, with negative medium- and long-term consequences for agricultural productivity and food security (Pandey *et al.*, 2007).

Impact of disasters on natural resources and ecosystems that sustain agriculture

Agricultural production relies on the availability and quality of natural resources and ecosystem services. Population growth, climate change and unsustainable management are among the key factors posing a threat to land, water and biodiversity, which form the natural base of agriculture. In turn, the depletion and degradation of natural resources increase the vulnerability and exposure of farmers to natural hazards, leading to more frequent and more harmful disasters. Deforestation, water resources depletion, land degradation, desertification, and degradation of coastal ecosystems such as mangroves and corals, all reduce nature's capacity to defend itself against natural hazards, and aggravate the impact of disasters (FAO, 2013b).

Disasters also have direct and indirect negative consequences on the natural resources and ecosystems that sustain agriculture. These might include, among others, surface and groundwater depletion and contamination, increased soil erosion, damage to native forests, mangroves, wetlands, salinization of soils, damage to coral reefs, and biodiversity loss. Furthermore, the displacement of affected people in the aftermath of disasters could indirectly lead to increased pressure on natural resources (e.g. exploitation of forest and water resources) in the areas surrounding displacement camps.

On average, each disaster caused about USD 32 million in damage and loss to the environment

A review of 74 PDNAs conducted in developing countries between 2006 and 2016 revealed that each disaster caused on average about USD 32 million in damage and loss to the environment, though in most cases that figure is likely an underestimate.

Drought is a major cause of water shortage and soil erosion and has devastating impacts, especially in countries with reduced capacity to absorb the shocks. For instance, in the Marshall Islands, the El Niño-driven drought in 2015 and 2016 led to the depletion of the already scarce water resources in storage facilities, combined with increased salinity of groundwater to unsafe levels (Government of Marshall Islands, 2017). In arid and semi-arid areas, prolonged or frequent episodes of drought can lead to the irreversible stage of desertification unless prevention measures are adopted.

Floods are frequently associated with water contamination and accelerated processes of soil degradation. When water recedes after flooding, some of the pollutants in the water are left in the soil. Silt and contaminated water degrade soils, particularly in cultivated areas. For instance, the floods that affected Sri Lanka in 2016 caused soil erosion and accumulation of silt in low agricultural lands, as well as water contamination in dug wells, causing widespread negative impacts on agricultural production that were estimated at about USD 2.6 million in damage (Government of Sri Lanka, 2016).

Tropical storms can also cause environmental damage over large areas by damaging natural resources and ecosystems that provide essential services for agricultural production. Tropical Cyclone Pam, which hit Vanuatu in 2015, caused severe damage to marine and coastal ecosystems such as coral reefs and mangroves, and to tropical forests. This compromised important ecosystem services such as water regulation, nursery services for fisheries, and protection against natural hazards, estimated at about USD 49.3 million in damage (Government of Vanuatu, 2015).

Ethiopia 2010 ⟨ A farmer woman feeding cattle with Multi-Nutrient Blocks

Chapter II
Impact of natural disasters on crop and livestock production in developing countries

One of the most direct ways in which disasters affect agriculture is through reduced production. This results in direct economic loss to farmers that can cascade along the entire value chain, affecting agricultural growth and rural livelihoods. This chapter examines the extent of crop and livestock production loss due to natural disasters over the last decade. The cumulative effect of over 330 large-, medium- and small-scale disasters is examined and production loss over the entire crop and livestock commodity range is quantified to adduce a holistic estimation of the cost of natural disasters for agriculture in developing countries.

From the country-wide devastation of the 2010 earthquake in Haiti to the drought-stricken rural livelihoods of East Africa in the periodic grip of El Niño and the many Indian Ocean fishery systems affected by tsunamis and cyclones, the harmful effects of natural disasters prevail across continents, climates and sectors. **Heavily reliant on weather, climate and water for its ability to thrive, agriculture is particularly vulnerable to natural disasters.** Agricultural impacts of disasters can be multipronged and long-lasting such as the contamination of water bodies, loss of harvests or livestock, outbreaks of disease or destruction of irrigation systems and other infrastructure.

While detailed regional and local accounts of the effects of extreme weather events exist, usually – although not necessarily adequately – documented through a PDNA process, the global-scale effects of droughts, floods and extreme temperatures on agricultural production are yet to be quantified on a consistent and systematic basis.

FAO's 2015 report on *The Impact of Disasters on Agriculture and Food Security* revealed that between 2003 and 2013 developing countries suffered a total of USD 80 billion in crop and livestock production loss due to 140 large-scale disasters; 83 percent of this impact was caused by major floods and droughts. This disaster-triggered loss occurred in countries where agriculture is one of the main economic drivers, often contributing up to 30–40 percent of both national GDP and employment.

FAO's 2017 analysis covers 332 disasters in 87 developing countries across Africa, Latin America and the Caribbean, Asia and the Pacific Islands

This report looks at how trends in agricultural impacts of disasters have evolved, taking stock of both new developments and persisting tendencies. The extent and cost of reduced agricultural production due to disasters is examined for the 2005–2015 period. The scope and level of analysis extends beyond the large-scale disaster focus to include both medium- and smaller-scale disasters affecting over 100 000 people or 10 percent of the national population.[1] This allows for a special emphasis on SIDS. Sector economic loss from disasters is estimated by analysing long-term trends in crop and livestock production flows and associated deviations following disasters. The analysis covers 332 disasters in 87 developing countries across Africa, Latin America and the Caribbean, Asia and the Pacific Islands. The crop and livestock sectors are considered as a whole, looking at every reported commodity produced in each country (or an average of 114 commodities per country).[2]

Production loss – an overview

One of the most direct ways in which disasters impact agriculture is through reduced production. This results in direct economic loss to farmers, which can cascade along the entire value chain, affecting the growth of the entire sector or even economy. Reduced production therefore remains not only the most direct measure of disaster impact, but also a strong indication of the scope and scale of that impact.[3] **Between 2005 and 2015, approximately USD 96 billion was lost as a result of declines in crop and livestock production in developing countries following natural disasters.**

---

1 The 2015 report only considered disasters affecting over 250 000 people. Because the 2017 report considers disasters affecting over 100 000 people or 10 percent of the national population, it focuses equally on large-, medium- and smaller-scale disasters, including those affecting SIDS, which have lower population levels.
2 The 2015 report estimated loss for main crop commodities. The 2017 report analyses the overall crop and livestock sectors, considering all reported commodities.
3 A detailed explanation of the parameters of the analysis and the calculations behind the production loss figures is provided at the end of this chapter.

Figure 1. Total production loss, 2005–2015 (in USD billion)

In Africa, loss over the ten-year period amounts to USD 26 billion; in Latin America and the Caribbean USD 22 billion, while in Asia cumulative loss amounts to USD 48 billion

For Africa (both sub-Saharan and North Africa), loss over the ten-year period amounted to USD 26 billion, for Latin America and the Caribbean USD 22 billion, while in Asia cumulative loss amounted to a staggering USD 48 billion, making up 50 percent of total loss. These were mostly accrued in Southern Asia (USD 32 billion) and Southeast Asia (USD 14.5 billion). The loss estimated for Oceania is much lower in absolute terms, at USD 4 million.

Loss as share of potential production

The extent of disasters in agriculture is even more evident when loss is measured as percentage of potential production (Figure 2). This is computed here as the difference between actual and expected production in the disaster years. The expected production is the amount that would have materialized in the absence of the hazardous events.

Despite the smaller global scale, several African regions – particularly Central and Southern – show high production loss in these terms, along with Western Asia, the Caribbean and Polynesia. In these regions, disasters levy a toll of about 8–10 percent on potential production in disaster years. SIDS present a particular case in point: while their loss was relatively low in absolute terms, it constituted a large burden on the local agricultural sector, destroying 7 percent of potential production in Polynesia and the overall group of Pacific SIDS, and 9 percent in the Caribbean. On a global scale, loss from natural disasters accounts for about 4 percent of potential production, which is a significant amount. **Production disruptions of that magnitude can have severe impacts on international markets and affect global food supply.**

Figure 2. Production loss due to natural disasters as percentage of potential production, by region, 2005–2015

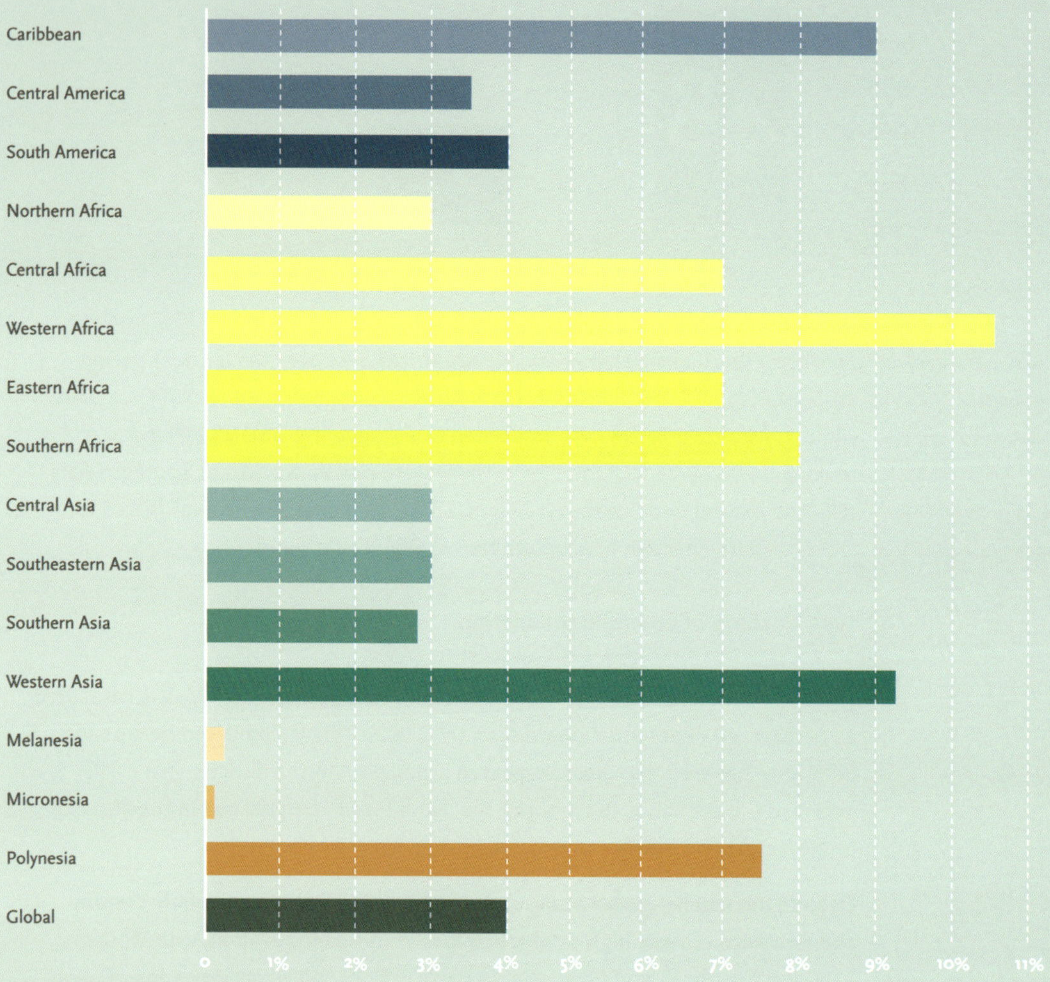

Caribbean	
Central America	
South America	
Northern Africa	
Central Africa	
Western Africa	
Eastern Africa	
Southern Africa	
Central Asia	
Southeastern Asia	
Southern Asia	
Western Asia	
Melanesia	
Micronesia	
Polynesia	
Global	

0 1% 2% 3% 4% 5% 6% 7% 8% 9% 10% 11%

Figure 3. Total production loss per disaster type, 2005–2015

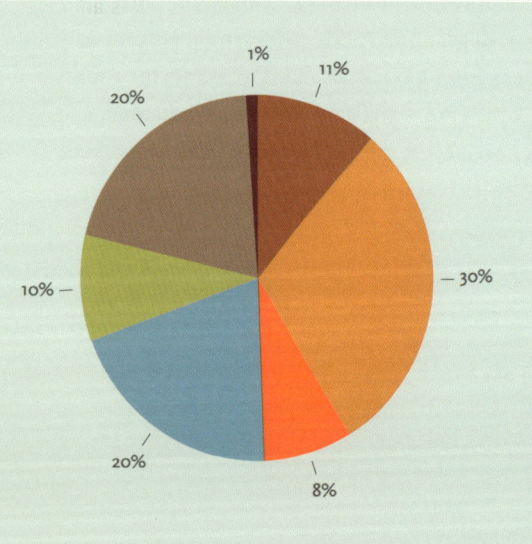

1% 11% 20% 30% 10% 20% 8%

Figure 4. Production loss by region and per disaster, 2005–2015 (in USD billion)

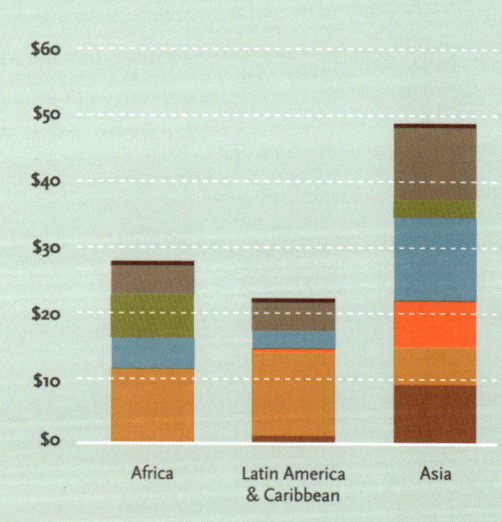

$60
$50
$40
$30
$20
$10
$0

Africa   Latin America & Caribbean   Asia

Legend: **Earthquakes/landslides/mass movements,** Drought, **Extreme temperatures,** Floods, **Crop pests/animal diseases/infestations, Storms, Wildfires.**

Different disasters, different outcomes

In order to better inform DRR policy, it is important to develop an understanding of which disasters and disaster types have the largest impact on agriculture and are responsible for the greatest loss. Between 2005 and 2015, in developing countries:

→ **floods** caused 20 percent of the cumulative production loss in crops and livestock, amounting to just under USD 19.5 billion;

→ **drought** caused 30 percent of agricultural loss, which amounted to over USD 29 billion;

→ other meteorological disasters, such as **extreme temperatures and storms,** set the sector back over USD 26.5 billion, representing 28 percent of overall production loss;

→ biological disasters, such as **diseases and infestations,** accounted for around 10 percent of total loss (or approximately USD 9.5 billion);

→ **wildfires** were relatively less impactful accounting for a moderate share of 1 percent of total loss, or just under USD 1 billion.

Figure 4 shows that drought accounted for the majority of loss in Africa and Latin America. In Asia, floods and storms were the disasters mostly responsible for reduced agricultural production. Crop pests and animal diseases are among the costliest disasters in Africa – more so than in Latin America and Asia – accounting for over USD 6 billion in agricultural loss between 2005 and 2015. Alongside floods, Asian agricultural systems are equally confronted with earthquakes/ tsunamis and extreme temperatures, which account for over USD 9 billion and USD 7 billion of loss, respectively. **Drought appears to be the most expensive disaster in Latin America and Africa, where the resulting crop and livestock loss amounted to USD 13 billion and USD 10.7 billion respectively.** At 1 percent, the significance of wildfires appears to be relatively small. Though it is safe to assume they have a negligible effect on crop and livestock production, wildfires are among the main disasters affecting the forestry sector worldwide. This report takes an in-depth look at the impact of disasters on forestry, including wildfires (Chapter 6).

Crop and livestock production loss – trends around the world

On a year-by-year basis over the last decade, disasterous events have inflicted a consistently high loss on crop and livestock production in developing countries. **In five out of the last ten years, loss was estimated to be higher than USD 10 billion per year, and the overall trend points to an increase.**

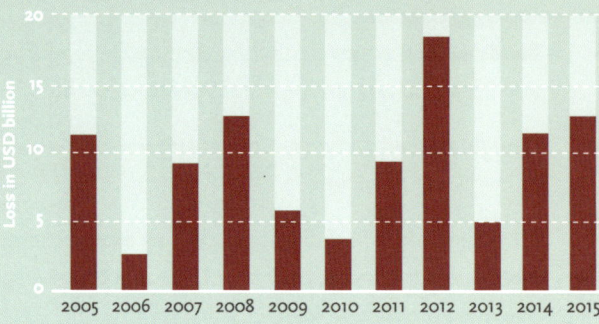

Total loss in crop and livestock production, developing countries in all three regions

Figure 6. Total loss in crop and livestock production due to natural disasters – Africa.

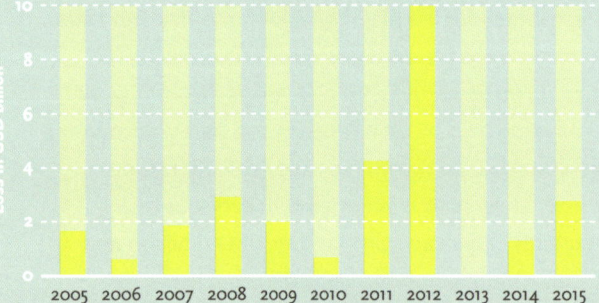

In Africa – including both the sub-Saharan and the North regions – loss has fluctuated widely, with peaks in 2011 and 2012, mostly driven by drought in the Sahel and Horn regions.

Figure 7. Total loss in crop and livestock production due to natural disasters – developing countries in Latin America & Caribbean.

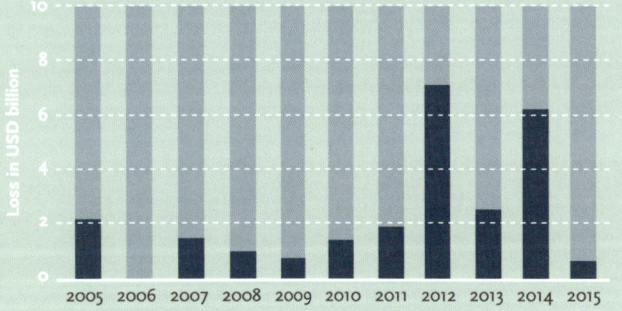

In Latin America and the Caribbean, loss has increased considerably over the past five years. The pronounced peaks in 2012 and 2014 in particular reflect severe La Niña-related drought episodes, which ravaged crop harvests in Argentina and Brazil in 2012 and much of Central America in 2014, especially the crop and livestock sectors in El Salvador, Guatemala and Honduras.

Figure 8. Total loss in crop and livestock production due to natural disasters – developing countries in Asia.

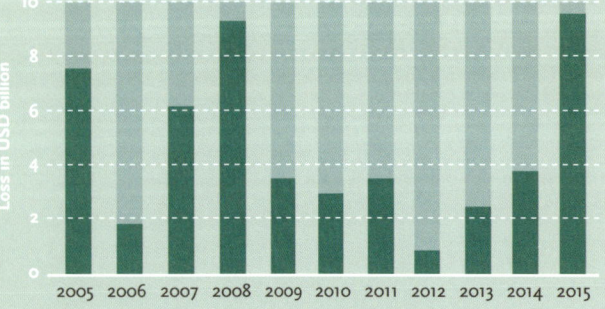

In Asia on the other hand, the overall level of loss in agricultural production is comparatively higher, with peaks in 2008 and 2015. This increase was observed mainly in Southern Asia and attributed to the series of monsoon floods and earthquakes reported at the time.

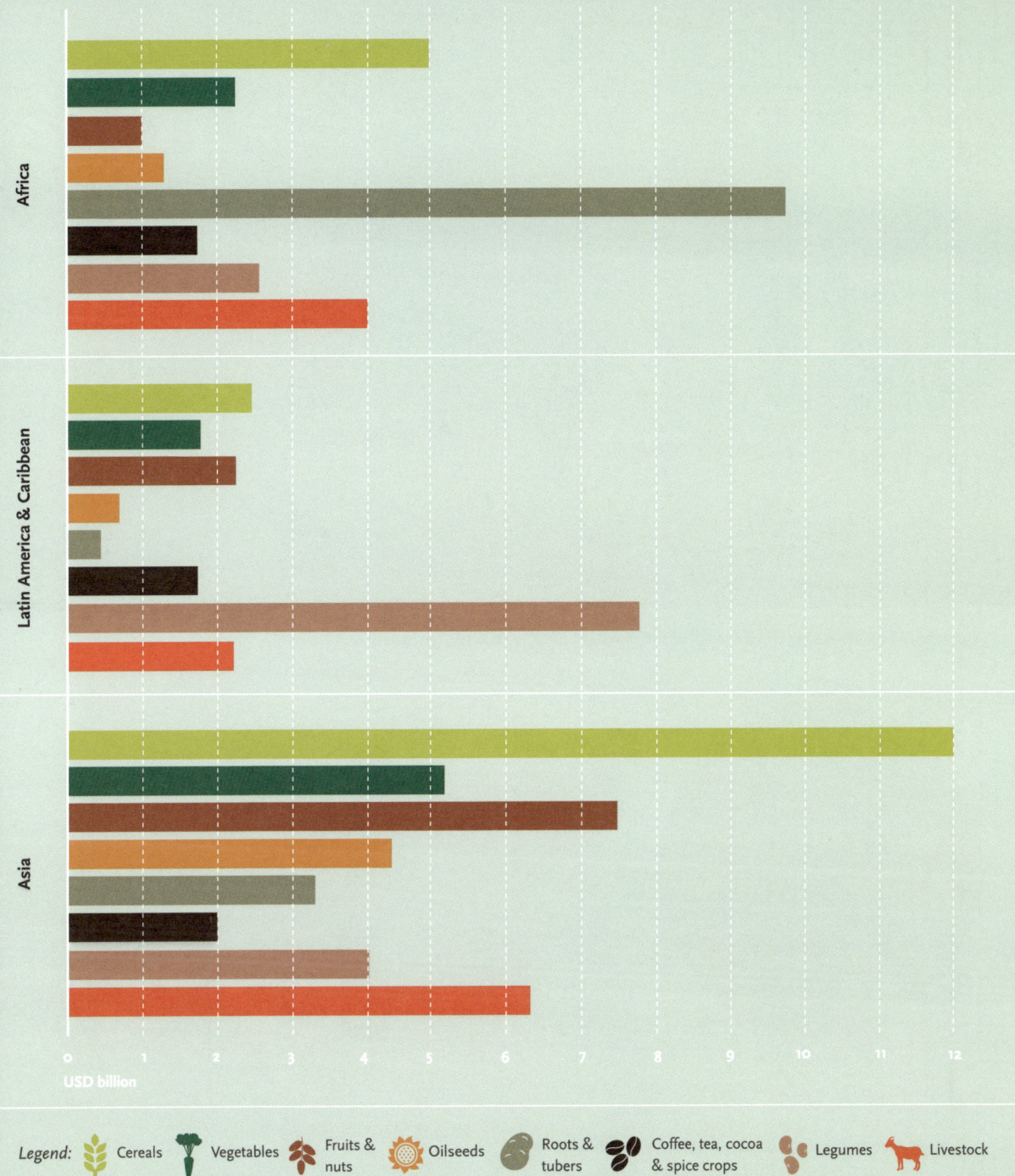

Figure 9. Production loss by commodity group, 2005–2015 (in USD billion)

USD billion

Legend: Cereals | Vegetables | Fruits & nuts | Oilseeds | Roots & tubers | Coffee, tea, cocoa & spice crops | Legumes | Livestock

Not all commodities are affected equally by disasters across regions (Figure 9). The distribution of impact across commodity groups largely reflects their relative importance in the production mix of different areas. Therefore, over the past ten years the production of roots and tubers – such as potatoes, sweet potatoes, cassava and yams – sustained the highest loss in Africa, amounting to just over USD 9 billion. Cereal and livestock production loss followed closely at USD 5 billion and USD 4 billion, respectively. In Asia, disaster-related production loss was high across all commodity groups. However cereal production stands out with a staggering cumulative loss of about USD 12 billion over the past decade. Rice and wheat were among the commodities most affected. Furthermore, disasters in Asia also had a serious impact on fruit and nut production (loss of USD 7.3 billion), livestock production (loss of just over USD 6 billion) and vegetable production (loss of about USD 5 billion). On the other hand, disasters striking developing countries across Latin America and the Caribbean mostly affected the production of leguminous crops such as beans, lentils and chickpeas, causing a loss of just under USD 8 billion between 2005 and 2015.

Figure 10. Average absolute and relative impact of disasters in SIDS and NON-SIDS countries (2006–2016)

Population affected	Damage and loss in agriculture

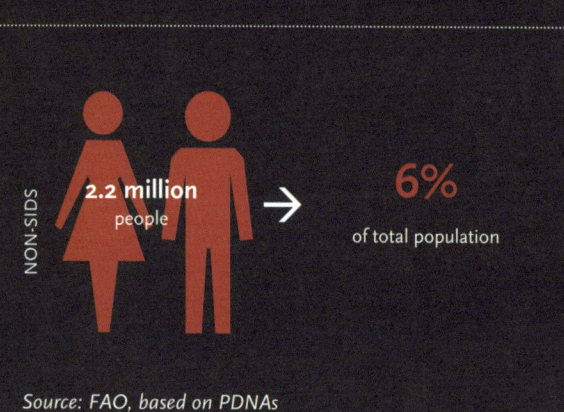

SIDS

440 000 people → **18%** of total population

$50 million → **19%** of agriculture value added

NON-SIDS

2.2 million people → **6%** of total population

$580 million → **8%** of agriculture value added

Source: FAO, based on PDNAs

Figure 11. All SIDS – Crops and livestock production loss per disaster type (2005–2015)

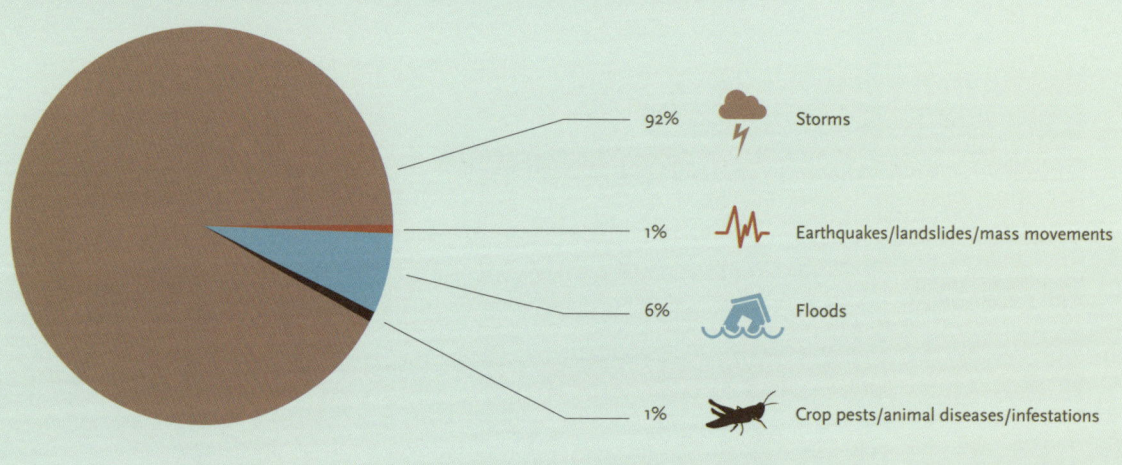

92% Storms

1% Earthquakes/landslides/mass movements

6% Floods

1% Crop pests/animal diseases/infestations

Figure 12. Loss from natural disasters in Pacific SIDS – overview by disaster (2005–2015)

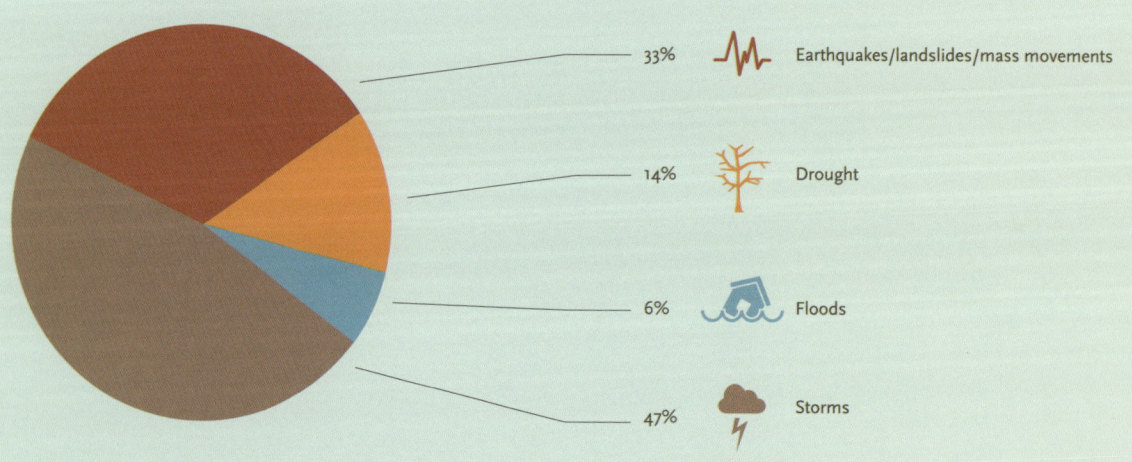

33% Earthquakes/landslides/mass movements

14% Drought

6% Floods

47% Storms

A focus on Small Island Developing States (SIDS)

Given their remote location in the Pacific and Indian Oceans, the Caribbean and along the coastline of Africa, SIDS are particularly vulnerable to the detrimental consequences of natural disasters. They suffer disproportionately from events such as tsunamis, earthquakes, storms and floods, which affect their economies, territories, and at times threaten their very existence. **Over the past decade, the economic loss from disasters in SIDS showed a staggering increase from USD 8.7 billion (between 2000–2007) to over USD 14 billion (between 2008–2015).[4] Climate change poses a further challenge as rising sea levels are responsible for extreme coastal flooding, endangering the livelihoods of over 4.2 million people.**

Damage and loss caused by a single disaster in agriculture corresponds to about 19 percent of the agricultural value added in SIDS, compared to only 8 percent in non-SIDS countries

The agricultural damage and loss attributed to disasters in SIDS over the past decade has been significant. In absolute terms, the cost of disasters for the agricultural sector is lower in SIDS than in non-SIDS (about 10 times lower). In relative terms, however, agricultural damage and loss in SIDS represents a substantial share of the sector's GDP. On average, the damage and loss caused by a single disaster in agriculture corresponds to about 19 percent of the agricultural value added in SIDS, compared to only 8 percent in non-SIDS countries.

Furthermore, disasters in SIDS affect a larger share of the population: on average, 18 percent of the total population is affected after each disaster in SIDS, compared to 6 percent in non-SIDS countries. **Disasters' relatively large economic impact on agriculture in SIDS, combined with the lack of coping capacity and high exposure and vulnerability, often create trade-offs between disaster recovery spending and budgetary allocations to development priorities**, thereby compromising future sectoral and overall growth.

Tropical Cyclone Winston caused about USD 100 million in damage and loss to fisheries

The most prominent disasters to hit SIDS are storms (up to 92 percent of impact), followed by floods, tsunamis, earthquakes and landslides. Tropical Cyclone Winston, which hit Fiji in February 2016, caused about USD 100 million in damage and loss to fisheries, corresponding to about 2.3 percent of the country's GDP in 2015.

The picture changes when a regional focus is introduced. The breakdown of loss by disaster for Pacific SIDS reveals that both storms and earthquakes and/or tsunamis account for the greatest part of all agricultural loss. Drought and floods are also detrimental disasters, however their share of total loss is relatively smaller compared to other geographical regions.

---

4 OECD & The World Bank, 2016.

Drought loss as a percentage of potential production, 2005–2015

Niger
4.6%

Chad
4.2%

Mali
10.4%

Ethiopia
3.6%

Somalia
2.3%

Senegal
2.5%

Uganda
3.9%

Kenya
3.0%

Burundi
4.8%

United Republic
of Tanzania
3.5%

Mozambique
20.1%

Malawi
9.3%

Legend / drought occurrence : ● 1-2 ● 3-5 ● 6-7

Drought occurrence data is based on EM-DAT CRED, including all reported drought occurrences between 2005 and 2015. Agricultural loss from drought, expressed as a percentage of potential production, is calculated based on FAOSTAT production data for droughts affecting over 100 000 people or 10 percent of the national population.

Drought at the core

Drought continues to constitute a major challenge for agricultural systems across the Sahel, the Horn and part of the eastern regions of Africa. The extreme vulnerability to rainfall variability in the arid and semi-arid areas of the continent and the poor capacity of many soils to retain moisture result in an often devastating impact on the sector. Between 2005 and 2015 droughts were frequent and severe in many African countries (map below): there were 84 reported drought occurrences in 30 countries, which led on average to a loss of 3–4 percent from potential agricultural production, a number that can rise to 10 or even 20 percent in certain cases.

The severity of the economic impact of drought is growing. **Between 2005 and 2015, average annual loss in crop and livestock production in developing countries has skyrocketed**: from under USD 2 billion per year until 2010, to close to USD 8 billion in 2014 (Figure 13).

Despite the importance of this loss, a persistent limited availability of data and information is hampering proper understanding of the economic consequences of drought in the agricultural sector. **Recurrent droughts can lead to poor soil fertility, reduced output, loss of livestock, limited access to markets and a host of other constraints faced by smallholder farmers.** This makes agriculture a high-risk endeavour and can stifle investment, pushing the country into a cycle of underproduction, low income and persistent poverty. Therefore, the case for investing in resilience and drought risk reduction, including in data and information generation, is very strong.

Turn to Chapter 3 for a more detailed account of recent droughts in Ethiopia and their cost to the local and national crop and livestock sector.

Figure 13. Loss in crop and livestock production caused by drought – developing countries in Asia, Africa, Latin America & Caribbean, 2005–2015

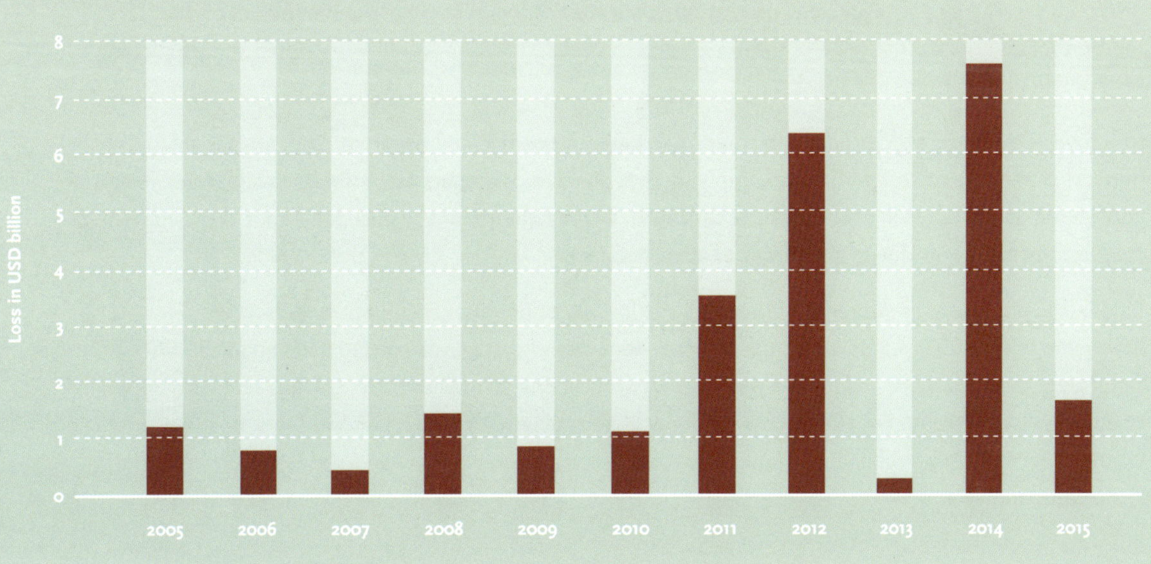

Calculating production loss – approach and analysis

Economic loss by sector from disasters for 2005–2015 is estimated by analysing long-term trends in crop and livestock production flows and associated deviations following disasters that occurred over the ten-year period. The analysis covers 332 disasters occurring in 87 developing countries across Africa, Latin America and the Caribbean, Asia and the Pacific Islands. Furthermore, the analysis considers the crop and livestock sector as a whole, looking at every reported commodity produced in each country (an average of 114 commodities per country). Finally, both large- and medium- (to small-) scale disasters are considered. Hazardous events considered are those that have affected 100 000 people or more, or at least 10 percent of the national population.[5]

It is important to underline that using deviations from trends in production as estimates of production loss implies rather strong assumptions and several limitations. Agricultural production is subject to significant year-to-year variability for many different reasons that are unrelated to the occurrence of disasters. By and large, annual production of each commodity can vary due to market trends and expected demand, normal climate variability, disease outbreaks or other immediate reasons at regional, national or local level. The use of "expected" production, as a starting point to measure the impact of disasters on production implies that none of these non-disaster related factors would have significantly affected production in the absence of a disaster.

Moreover, **deviations from production trends can be both positive and negative. Both are assumed as impacts of the disaster on production, implying that positive deviations from trends in a disaster year are considered as increases in production, occurring as a consequence of the disaster.** Again, such increases may in principle occur due to a plethora of other reasons.

Finally, the procedure employed assumes that the impact of the disaster on production is entirely exhausted in the same year in which the disaster occurs, and disregards cumulative impacts that may occur over more than one year. While this assumption is consistent with the emphasis on loss as opposed to damage (see Introduction, Key terms), it can still be problematic for certain products, such as perennial crops. Despite such possible limitations, this approach represents a good and viable option to run large-scale comparative assessments in the absence of more accurate data.

---

5   Similar estimates presented in the 2015 report considered as "disasters" the events that affected at least 250 000 persons and at least 25 percent of the population.

General parameters of the procedure

The impact of disasters for the 2005–2015 period is considered, while production trends are estimated based on data since 1980. The analysis is conducted for developing countries by region (Africa, Latin America and the Caribbean, Asia and the Pacific Islands) and for all reported commodities (in terms of yields), including all crops and livestock products reported in FAOSTAT, or 114 commodities on average per country. **The main data source on production and prices is FAOSTAT, while data on disaster occurrence and people affected is sourced from the EM-DAT CRED online database.** All main types of natural disasters are considered, including geophysical (earthquakes, tsunamis, mass movements), climatological (droughts, wildfires), meteorological (storms, extreme temperatures) and hydrological (floods).

Production trend estimates and loss calculations

A first threshold is applied to select only disasters affecting over 100 000 people and/or 10 percent of the population. A second threshold is used when in any given country disasters occur in two or more consecutive years. In this case, the average number of people affected in each of these disaster occurrences is calculated and only those disasters whose effect exceeds the resulting average are considered.

Following the establishment of relevant thresholds, the analysis proceeds with estimates of long-term production trends – linear, polynomial second order, polynomial third order, logarithmic – for the production (yields) of every commodity in every country for the period 1980–2015. A goodness-of-fit test is applied to determine and select the model that predicts actual production most adequately for each commodity.

Deviations from trends in the years in which a disaster occurred, for each country and commodity, are assumed to constitute estimates of production loss. To aggregate across products, production loss is converted in monetary terms using FAOSTAT prices of the year preceding the disaster in constant dollar value.

PART II Estimating damage and loss:
getting it right

Philippines 2013 〈 Typhoon Haiyan

Chapter III
Applying the methodology – trials from the Philippines and Ethiopia

Reducing disaster impact is possible only on the basis of accurate information and a thorough understanding of the impact. The methodology FAO has developed to assess the direct economic impact of disasters on agriculture will serve to monitor progress in achieving the targets of the post-2015 international resilience agendas, including the Sendai Framework and the SDGs. This chapter presents, applies and calibrates the foundations of that methodology through analysis of two contrasting disasters: the rapid-onset Typhoon Haiyan in the Philippines (2013) and the slow-onset drought in Ethiopia (2008-2011).

Towards a general assessment framework

Be it a major headlining catastrophe such as Typhoon Haiyan in the Philippines, or a localized event such as a flood in Bolivia that never made it into international news, natural hazards and disasters are sprouting up at an accelerated rate. Even during the preparation of this report, the count of events kept growing as hurricanes Irma and Maria devastated the Caribbean, a deadly 7.1-magnitude earthquake rocked central Mexico and monsoon rains caused severe floods in Myanmar, displacing thousands of people. The developed world is not spared either, as hurricanes Harvey and Irma struck the US mainland, causing damage in excess of USD 300 billion.

As evidenced throughout this report, extreme weather, climate and geophysical events have occurred with increasingly high frequency and magnitude over the past decade. Chapter 2 shows the cumulative impact of disasters on agricultural production, resulting in significant loss to the crop and livestock sectors. While the analysis exposes the great challenges posed by the general incidence of disasters, the particular consequences of individual events for the sector and its subsectors remain poorly analysed and largely under-reported. There is limited availability of agriculture-specific, systematic estimates of damage and loss following disasters, particularly those of smaller- and medium-scale. To address this structural gap, FAO has developed a standardized methodology to provide a set of procedural and computational steps for consistent damage and loss assessment across disasters and countries. It is grounded in and builds upon existing frameworks, tools and methods for disaster impact assessment, such as ECLAC's damage and loss assessment methodology (DaLA) and the PDNA methodology, while aiming to systematize and standardize the process at the global, national and local levels.

There is limited availability of agriculture-specific systematic estimates of damage and loss following disasters, particularly of smaller- and medium-scale

Structure

In order to capture the full impact of disasters on the agriculture sector, FAO's methodology for **damage** and **loss** assessment distinguishes between damage, i.e. total or partial destruction of physical assets, and loss, i.e. changes in economic flows arising from a disaster. Furthermore, each subsector is divided into two main components: **production** and **assets**. This allows for an estimation of the extent and value of damage and loss for all components in each subsector and for the formulation of a globally standardized assessment of the impact.

The **production component** measures disaster impact on agricultural inputs and outputs. Damage includes the value of stored inputs (e.g. seeds) and outputs (e.g. crops) that were fully or partially destroyed by the disaster. On the other hand, production loss refers to declines in the value of agricultural production resulting from the disaster. The **assets component** measures disaster impact on facilities, machinery, tools, and key infrastructure related to agricultural production. The monetary value of (fully or partially) damaged assets is calculated using the replacement or repair/rehabilitation cost, and is accounted for under damage. The Annex of this report provides a detailed description of the technical aspects of the methodology used for each sector, including an indication of the items and economic flows that should be considered in the assessment, as well as the proposed calculation methods for assigning monetary values to each category.

The monetary value of damaged assets is calculated using the replacement or repair/rehabilitation cost, and is accounted for under damage

Prospects

The Sendai Framework offers the opportunity to scale up DRR efforts in agriculture

Recently integrated into global resilience initiatives such as the Sendai Framework and the SDG agenda, **this methodology will further serve to measure progress towards reducing the monetary impact of disasters on agriculture.** The Sendai Framework offers the opportunity to scale up DRR efforts in agriculture, which can be measured against tailored development outcomes and calls for a more proactive and holistic approach to DRR. Through the dedicated indicator on direct agricultural loss attributed to disasters (Sendai Framework indicator C-2), FAO's new methodology has the opportunity to contribute to the first global system for recording disaster loss. Therefore, **it is important to demonstrate that it constitutes a reliable, holistic and universal tool across all agricultural sectors** (crops, livestock, forestry, fisheries and aquaculture) **for a varied range of disasters, and one that accommodates the various levels of data availability**.

Before FAO's methodology can be more widely applied as the single framework for damage and loss assessment in agriculture, it must be put to the test. This chapter takes a first step toward that end, employing the new methodology to quantify the impact of two very different disasters in two very different contexts: the sudden-impact event of Typhoon Haiyan in the Philippines, and the slow-onset disaster of drought in Ethiopia.

Typhoon Haiyan in the Philippines

Typhoon Haiyan hit the central Philippines on 8 November 2013. Its winds reached more than 300 km per hour, the strongest wind speed ever recorded in the country during a cyclone landfall (Takagi & Esteban, 2016). Haiyan's storm surges reached 5.3 metres, causing widespread devastation and loss of life (Lagmay *et al*, 2014). At least 6 300 people died in the cyclone and its immediate aftermath, which affected an estimated 16 million people, and damaged or destroyed more than 1.1 million homes, as well as public infrastructure and agricultural land across 41 provinces, as reported by the National Disaster Risk Reduction and Management Council (NDRRMC, 2013).

Although Typhoon Haiyan struck after the harvest, sparing the country even greater devastation, it nevertheless caused extensive damage to the agriculture sector, especially in areas heavily dependent on crop production and fishing. **Hundreds of thousands of hectares of rice and other key crops were affected**. Rural infrastructure and storage were also severely damaged. The storm surge wiped out fishing communities and destroyed boats and gear. Over one million fishermen and farmers were estimated to be in need of urgent assistance to restore their livelihoods and productive assets.

The monetary impact of Typhoon Haiyan on the Philippines' agriculture sector was retroactively calculated using FAO's methodology

The monetary impact of Typhoon Haiyan on the Philippines' agriculture sector was retroactively calculated using FAO's methodology on damage and loss assessment and its associated computation methods (Annex). The retroactive assessment was carried out at provincial level, using primary data on physical damage from post-disaster impact assessments conducted by the Government. Gaps in primary data were addressed through estimation procedures using secondary information.

Table 1. Main components of the agriculture sector in the Philippines

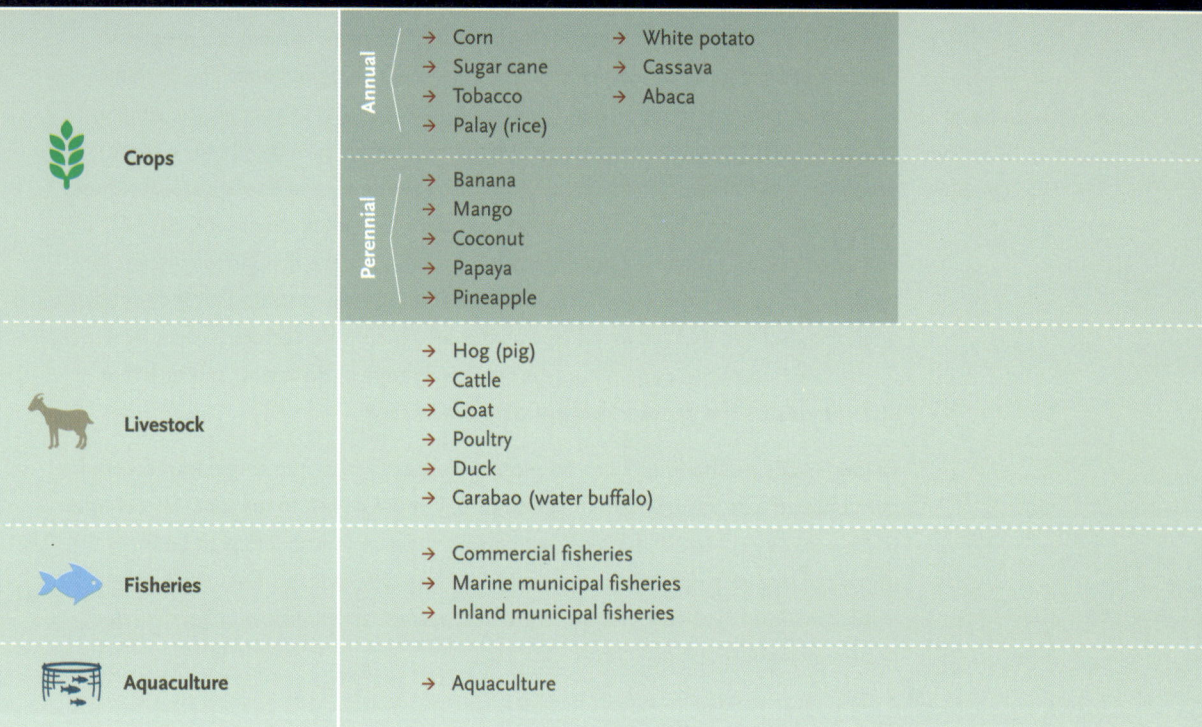

Crops	**Annual** → Corn → White potato → Sugar cane → Cassava → Tobacco → Abaca → Palay (rice)	
	Perennial → Banana → Mango → Coconut → Papaya → Pineapple	
Livestock	→ Hog (pig) → Cattle → Goat → Poultry → Duck → Carabao (water buffalo)	
Fisheries	→ Commercial fisheries → Marine municipal fisheries → Inland municipal fisheries	
Aquaculture	→ Aquaculture	

In order to conduct a damage and loss assessment, the components relevant to each subsector must be identified. Based on data available at national and subnational level, 22 components were selected and analysed to quantify the monetary value of damage and loss in crops (annual and perennial), livestock, fisheries (inland and marine fisheries), and aquaculture (Table 1).

Given that the analysis is largely based on a set of assumptions and externally-derived parameters, assessment results may be biased for a variety of reasons: a lack of data requires the use of estimates; errors may occur due to externalities or a lack of sensitivity in measurements; the knowledge-based features of the methodology itself may influence outcomes depending on the information source.

The margin of error in estimating the damage and loss caused by Typhoon Haiyan complements FAO's methodological approach and ensures the soundness of its findings

In order to represent at least part of this threefold variability in the outcome measurements and provide a margin of error for the results, a two-step error analysis is employed which considers the variability in defining the exogenous parameters. The first step defines a "min-max interval" for the external parameters of every component in Table 1. Based on various information sources, an average, a minimum and a maximum value is defined for each parameter. **The resulting values for damage and for loss are then calculated three times for each component using: 1) the average values of those exogenous parameters, 2) the values that minimize the outcome, and 3) the values that maximize the outcome.**

Tables 2 and 3a-d summarize the outcomes of FAO's methodological assessment, both in general and by subsector. Table 2 does so while noting the margin of error for each figure, towards which both the min-max and confidence intervals contribute. Indicating the minimum and maximum parameters for damage, loss, and for damage + loss ought to support a high level of confidence in FAO's methodology and promote its quick adoption.

Overall, the combined **damage and loss caused by Typhoon Haiyan to agriculture amounted to over USD 1.4 billion.** Most of the impact – over 74 percent of combined damage and loss – was borne by the crops sector, which lost over USD 1 billion in the typhoon's aftermath. Total damage and loss to perennial crops amounted to USD 857 million, of which the most affected crop was coconut (USD 688 million), followed closely by banana, mango, papaya and pineapple. According to Government data, a total of 441 257 hectares of coconut plantations were affected by the typhoon, 40 percent of which were fully damaged with no chance of recovery. The high value of coconut plantations, combined with the long time required for new coconut trees to become fully productive again, caused extensive setbacks in agricultural value added.

Total damage and loss in perennial crops amounted to USD 857 million, of which the most affected crop was coconut (USD 688 million)

Table 2. Summary of damage and loss caused by Typhoon Haiyan on agriculture (USD million)						
	Damage	Margin of error for damage	Loss	Margin of error for loss	Total DL	Margin of error for DL
Crops	432	[387–477]	618	[522–704]	1049	[908–1181]
Production	388	[347–428]	618	[522–704]	1 005	[868–1132]
Assets	44	[40–49]	–	–	44	[40–49]
Livestock	37	[32–43]	57	[31–80]	94	[63–123]
Production	32	[27–37]	57	[31–80]	89	[58–117]
Assets	5	[4.7–5.7]	–	–	5	[4.7–6]
Fisheries/ Aquaculture	53	[47–58]	227	[204–250]	280	[252–308]
Production	–	–	227	[204–250]	227	[205–250]
Assets	53	[47–58]	–	–	53	[47–58]
Total	522	[423–526]	902	[757–1 034]	1 423	[1 223–1 612]

Table 3a. Annual crop damage and loss by commodity (USD million)

Crop	Production		Assets	Total
	Damage	Loss	Damage	
Abaca	0	7	1	8
Cassava	0	2	0.1	2
Corn	2	10	2	14
Paddy rice	6	55	6	67
Sugar cane	0	97	5	102
Tobacco	0	0.07	0	0.07
White potato	0	0.02	0	0.02
Total	8	171	14	193

Table 3b. Perennial crops damage and loss by commodity (USD million)

Crop	Production		Assets	Total
	Damage	Loss	Damage	
Banana	66	84	1.7	152
Coconut	309	350	28.6	688
Mango	3.6	11	0.06	14
Papaya	0.8	1	0	2
Pineapple	0.4	0.7	0	1
Total	380	446	30	856

Table 3c. Livestock damage and loss by commodity (USD million)

Animal	Production		Assets	Total
	Damage	Loss	Damage	
Cattle	1	5	0.2	6
Carabao	2	4	0.3	6
Chicken	3	6	0.4	9
Ducks	0	0	0	0
Hogs	26	40	4	70
Goat	0.5	2	0.1	2.6
Total	32.5	57	5	94

Table 3d. Fisheries damage and loss by commodity (USD million)

	Production		Assets	Total
	Damage	Loss	Damage	
Fisheries	n/a	145	34	179
Aquaculture	n/a	82	19	102
Total	n/a	227	53	280

Total damage and loss to annual crops amounted to USD 193 million, of which USD 171 million was in production loss, and USD 22 million in production and asset damage. The most affected annual crop was sugar cane, which sustained over USD 102 million in damage and loss, followed by rice with USD 67 million. Given the high commercial value of sugar cane (raw sugar was the third most important export commodity in 2011), the economic impact of Typhoon Haiyan on production had far-reaching economic consequences and significantly affected agricultural value added. Furthermore, the country's food security outlook was severely dented by the reduced rice production, since the majority of affected provinces are the largest rice producers in the country.

The most affected annual crop was sugar cane, which sustained over USD 102 million in damage and loss, followed by rice with USD 67 million

Fisheries and aquaculture constitute the second most-affected sector (20 percent of combined overall damage and loss), where impact amounted to a total of USD 280 million. The destruction of boats and other assets as a result of the strong winds and ocean surge had a significant impact on means of production, causing a decline in fish catch in many of the affected regions and provinces. Total **damage to assets** was around **USD 53 million.**

Finally, the monetary cost to the livestock sector was the smallest relatively at USD 94 million. Hog production accounted for the majority of impacts, followed by chicken, carabao, cattle and goats.

Comparing results for Typhoon Haiyan – how do we fare?
In the months following the typhoon, the Government of the Philippines conducted needs assessment studies with support from international organizations, including FAO. At that time, total damage and loss in agriculture was estimated at USD 1.4 billion – closely matching the 2017 results obtained through FAO's new methodology. The two assessments differ, however, with regard to monetary values assigned to the damage and loss categories. The earlier effort allocated an almost equal distribution between damage and loss – at USD 741 million and USD 733 million respectively. **The FAO approach reverses those categories. It concludes that 63 percent of the impact was loss and 37 percent damage. Despite this discrepancy, which mainly reflects differences in allocative choices, the overall compatibility of the assessments provides an all-important initial validation of FAO's methodology.** Further testing is needed to confirm this is the case universally, and to see if the methodology goes beyond that by facilitating damage and loss assessment in those situations where data is less available. Only then can it be considered a reliable framework for a holistic assessment of damage and loss in agriculture.

The earlier effort allocated an almost equal distribution between damage and loss; at USD 741 million and USD 733 million respectively. FAO's approach reverses those categories

Drought in Ethiopia

Radically different to the physical menace of strong winds and high wave surges, the slow-onset character of drought presents a particular challenge for any methodological approach. However, **the impact of drought on agriculture is significant and its quantification crucial for effective DRR policy.**
The number of drought-affected areas has grown in recent years, frequently compromising agricultural production, eroding livelihoods and triggering malnutrition and famine. While direct damage to agricultural assets and infrastructure is low (compared to destructions caused by typhoons, floods or earthquakes), **drought-related loss from crop failures and livestock mortality can be substantial. Drought can therefore be a prominent cause of food insecurity in developing countries.**

Its slow-onset nature, lack of visible physical damage, blurred temporal boundaries, and wide geographical reach make drought a particularly difficult hazard to assess with precision. Yet systematic assessment of drought-related impact on agriculture is crucial to inform evidence-based and cost-effective prevention and response strategies. This is especially relevant in countries where agriculture is of prime economic importance and where vulnerability to various shocks, weather abnormalities and climate change is particularly high. Ethiopia is a case at hand, regularly making international headlines with some of the most devastating droughts in recent decades.

When the La Niña- and El Niño-related hazards unfolded in Ethiopia in the latter part of the 2000s, they triggered the failure of several consecutive seasonal rains. This resulted in severe rain shortages across the southern highlands and the lowlands of the southeast in 2008, 2009 and 2010, culminating in the 2011 drought. The prolonged rainfall deficit resulted in water scarcity and depletion of groundwater reservoirs in the arid and semi-arid lowlands. Many of the affected areas reported consecutive poor harvests, production loss and poor livestock body conditions. The livestock mortality rate in the affected pastoral areas was 15–30 percent, with that of cattle and sheep as high as 40–60 percent in some areas (OCHA, 2011). The Ethiopian Government proclaimed an acute food crisis by July 2011 (Oxfam, 2012).

The livestock mortality rate in the affected pastoral areas was 15–30 percent, with that of cattle and sheep as high as 40–60 percent in some areas (OCHA, 2011)

The overall monetary value of agricultural impact caused by rain shortage and drought in 2008–2011 has not been estimated in national post-disaster assessments as only partial information is available through ad-hoc regional assessments. This case study therefore represents a first attempt to quantify the impact of drought on the Ethiopian agricultural sector. **Given the slow-onset nature of drought and the long-lasting span of its effects on the agricultural sector, this study analyses agricultural production trends from 2008 through 2011, thus taking into account both severe drought episodes and the lasting effects observed in the intermediary period.**

Damage and loss in the crop sector

The adverse weather conditions of 2008–2011 had a wide-ranging impact on crop production in the affected areas. The belg crop-producing parts of the country, where 80 percent of crop production depends on seasonal rains, experienced either poor harvests or complete crop failure. This disruption in agricultural activities during the main planting period significantly affected the area coverage and crop performance of both belg as well as meher crops such as maize and sorghum (FEWS NET, 2011).

Total damage and loss caused by drought in the crop sector in the meher seasons between 2008 and 2011 amounted to USD 37.3 million

By applying FAO's assessment methodology, crop damage and loss were calculated for the zones most affected by rainfall shortages over the period considered. Results of the assessment (Table 4) show that total damage and loss caused by drought in the crop sector in the meher seasons between 2008 and 2011 amounted to USD 27 million. Loss accounted for 96 percent of all impact, while damage contributed only 4 percent. The considerable difference in the amount of damage versus loss is due to the peculiar character of drought as a slow-onset phenomenon. Since no drought-induced physical damage to agricultural assets was reported (e.g. to irrigation systems, storage facilities, tools, machinery), the only damage considered in the analysis refers to the replacement value of fully damaged perennial crops (coffee) and the cost of additional inputs for replanting destroyed trees.

Figure 1 shows the distribution of damage and loss over the four drought years analysed. Among meher seasons, the majority of impacts occurred in 2011, mainly due to declines in cereal (especially teff and sorghum) in the Amhara and Southern Nations, Nationalities, and Peoples (SNNP) regions. The largest share of damage and loss can be traced to the impact of drought on teff production (Figure 2). The main teff-producing zones in the region were among the most affected, most notably North Shewa, West Shewa, Southwest Shewa and East Gojjam. Coffee plantations were significantly affected and key coffee-producing zones in the Oromia and SNNP regions (Illubabor, Kelem and Keffa) recorded a high degree of loss. Finally, the production of key staple crops such as sorghum and maize was strongly affected in Amhara and Dire Dawa regions, as well as in the Somali region, especially in Jijiga.

Table 4. Drought-related damage and loss in the Ethiopian crop sector, 2008–2011 (USD)							
	Damage		**Loss**		**Total Damage and Loss**		
Production	1 143 054	+	25 668 175	=	26 811 229		
Assets	–	+	–	=	–		
Total	**1 143 054**	+	**25 668 175**	=	**26 811 229**		

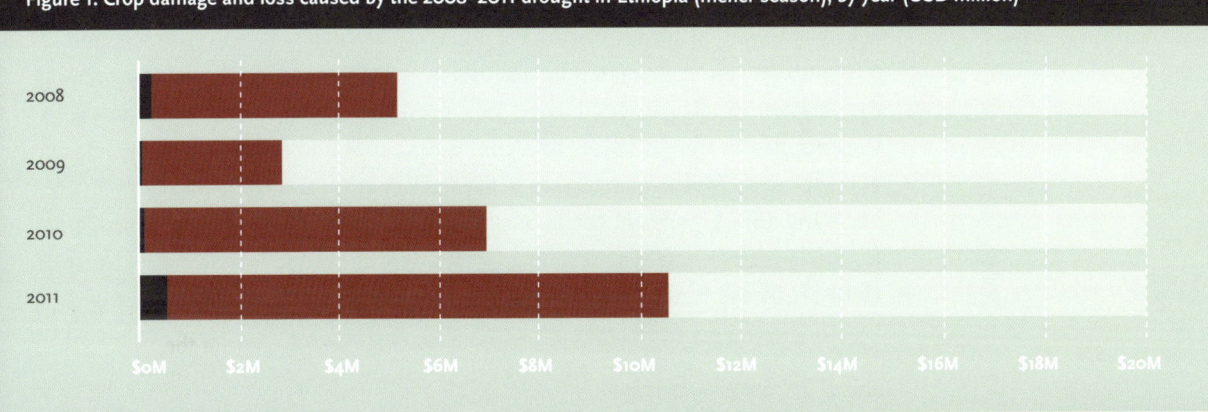

Figure 1. Crop damage and loss caused by the 2008–2011 drought in Ethiopia (meher season), by year (USD million)

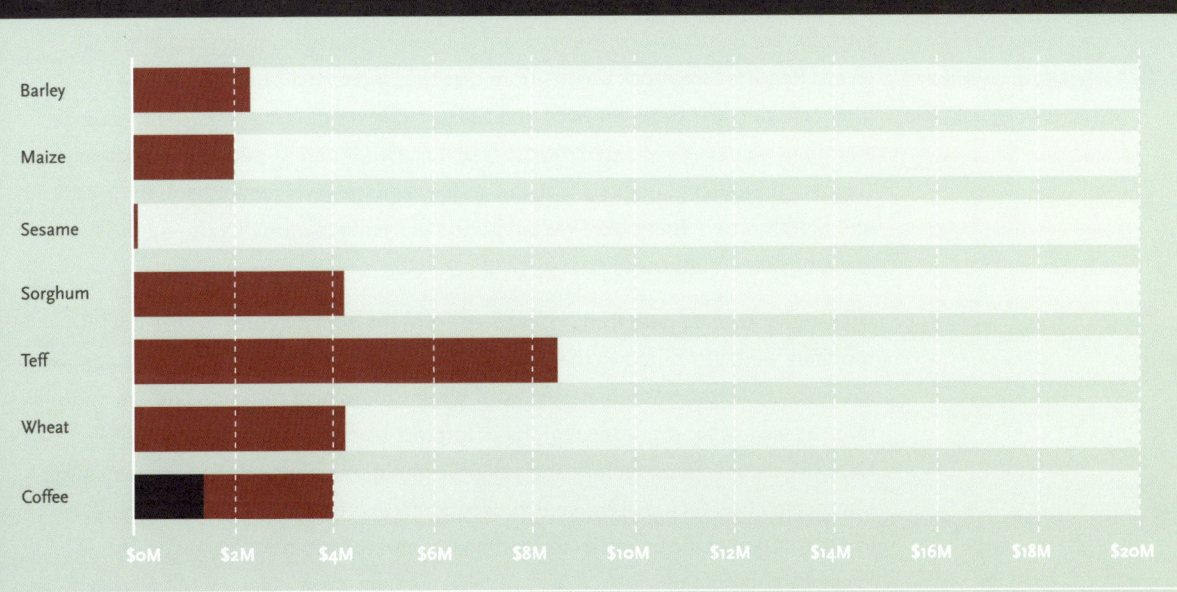

Figure 2. Crop damage and loss caused by the 2008–2011 drought in Ethiopia (meher season), by crop type (USD million)

Legend: **Damage,** Loss

Damage and loss in the livestock sector

The challenges that drought poses to pastoralists extend well beyond simple shortages of water and forage. The onset of drought is often associated with increased livestock death rates due to a fall in animal live-weight and increased susceptibility to disease. In the event of drought, the quality and quantity of both fodder and forage crops can significantly deteriorate, causing abrupt changes in livestock feeding patterns and nutritional status. As farmers find it more difficult to align production with their animals' nutritional requirements, livestock may succumb to starvation. This reduces animal immunity, making stocks more vulnerable to diseases likely to result in death. Livestock mortality can be both directly and indirectly associated with drought and extreme temperatures. The direct effects are related to rising temperatures and reduced moisture, creating heat stress and increasing the likelihood of morbidity and death. The indirect effects are related to reduced animal immunity, due to unfavourable climatic conditions and food scarcity.

While crop damage was substantial, the brunt of the drought was borne by the livestock sector, where pastoralist and agro-pastoralist livelihoods were devastated

Drought-associated mortality can vary across species and classes of stock, with cattle, horses, donkeys and sheep being generally less resistant than goats and camels. Age and condition can further determine vulnerability, making young animals, old stock and pregnant females within any given species particularly susceptible to disease outbreaks. The length, extent and severity of the drought episode also influence livestock mortality rates, which continue to rise as the period of nutritional stress lengthens and the degree of stress intensifies.

Estimating livestock loss due to drought in 2008 and 2011

The limited availability of data, constrained only to numbers of deaths per main livestock type, calls for estimating the impact of drought on the livestock sector based on deviations from trends in livestock mortality over the ten years preceding the disaster. To calculate production loss, livestock mortality numbers (in terms of livestock deaths from diseases and from other causes) were compiled in a dataset and analysed in terms of increases in livestock deaths for the two disaster years of 2008 and 2011. Significant deviations from linear mortality trends were considered for each livestock type. The resulting figures were converted into Tropical Livestock Units (TLU) and further standardized in order to obtain the monetary value of livestock loss. The results follow.

Trends in livestock mortality

Disease-associated mortality jumped to 52 percent in 2008 and to 23 percent in 2011, as compared to 9 percent in previous years

While livestock deaths – both caused by diseases and by other factors – have been increasing at regional and national levels since 2001, the reported mortality rates for 2008 and 2011 present a considerable increase from expected normal levels. **Reported livestock deaths from diseases increased by an average of 9 percent per year between 2001 and 2007.** However in 2008 and 2011 there was a jump in disease-associated mortality of 52 percent and 23 percent respectively (Figure 3). Similar changes are visible at the regional level for Amhara, Oromia and SNNP, where there were record jumps in livestock mortality during the two drought years (Figure 4).

Deviations from a linear trend computed from the above figures on livestock mortality were attributed to the drought conditions. The computation was applied to livestock deaths from diseases, from other causes and combined on the national level, as well as for the three regions of interest (Amhara, Oromia and SNNP). Using TLU conversion factors and weight estimations, these deviations indicating unexpected livestock deaths were subsequently expressed in monetary terms, in order quantify the extent of the monetary impact from the drought on the livestock subsector.

Figure 3. National overview – livestock deaths (millions of TLU)

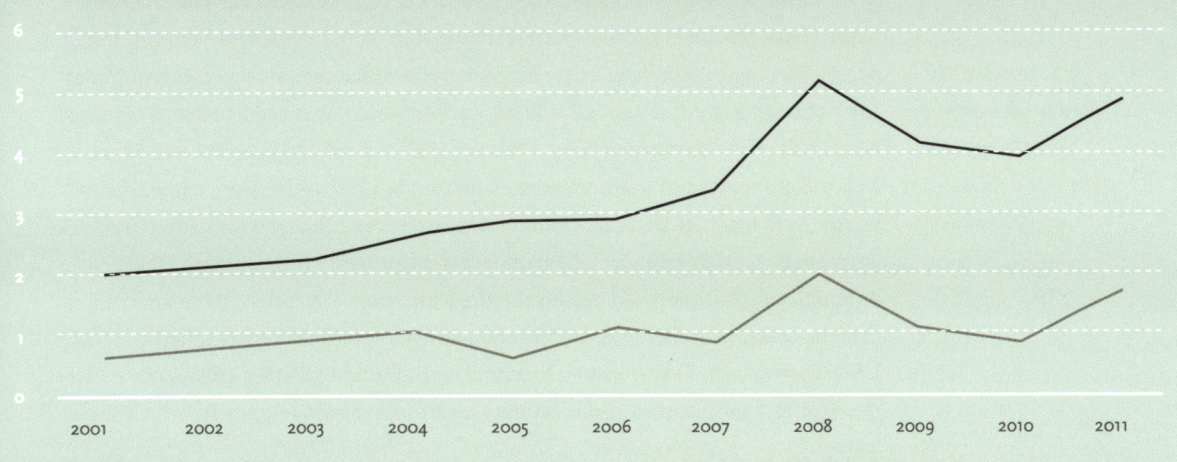

Legend: Deaths from disease, Deaths from other causes

Figure 4. Regional overview – livestock deaths caused by disease (millions of TLU)

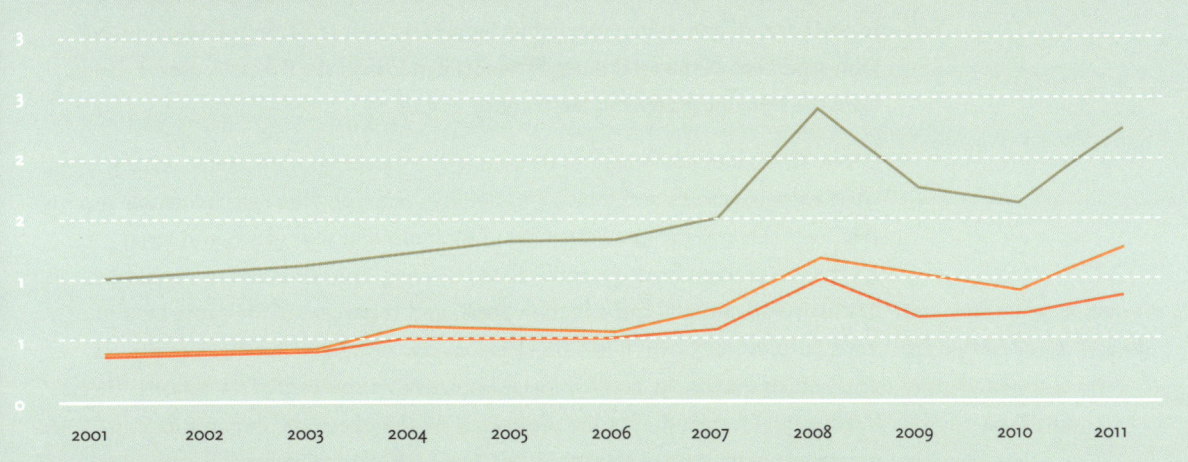

Legend: Amhara, Oromia, SNNP

Figure 5. Total livestock loss, country level (USD million)

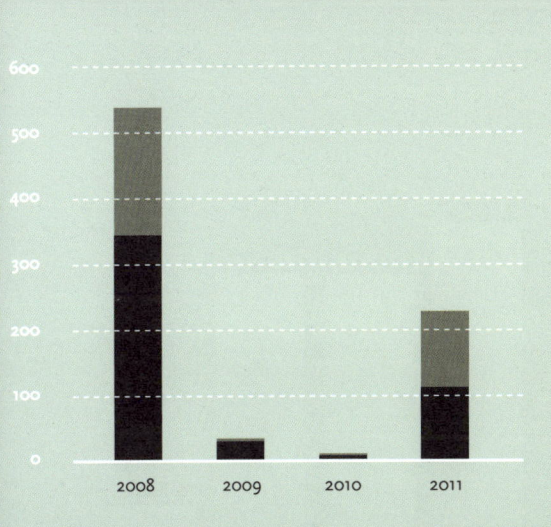

Legend: Livestock loss attributed to disease,
Livestock loss attributed to other causes

Figure 6. Total livestock loss by region, 2008–2011 (USD million)

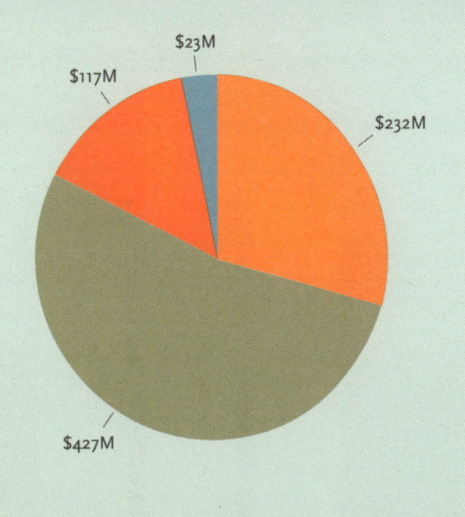

Legend: Amhara, Oromia, SNNP, Others.

The cost of drought to the Ethiopian livestock sector

In 2008, over 5 million heads of livestock were lost to drought-associated disease in Ethiopia, while over 2 million died of other drought-related causes. In Oromia alone these numbers exceed 1.5 million and 890 000, respectively. During the second drought episode of 2011, 6.5 million livestock died due to the drought at the national level, of which 4.8 million were reported as drought-related disease casualties. This loss amounts to a total of USD 757 million over the two years. In 2008 alone, loss in livestock was estimated at USD 535 million, while loss in 2011 was lower at USD 220 million. However, in both 2009 and 2010 livestock continued to exhibit abnormally high levels of mortality (beyond expected), although much lower than during the drought years. Considering the continued livestock deaths in this intermediary two-year period, total loss reached USD 800 million. This may be associated with the long-lasting effects of drought on the nutrition and immunity levels of animals (Figure 5).

In 2008 alone, loss in livestock was estimated at USD 535 million, while in 2011 it was lower at USD 220 million

At subnational level, an overwhelming majority of loss seems to originate from the three regions of Amhara, Oromia and SNNP. Oromia proves to have been the region most severely affected by the drought, with a cumulative loss amounting to USD 427 million between 2008 and 2011, representing 56 percent of all loss in the country. The livestock sector in Amhara was also highly impacted, accruing close to USD 232 million in loss, while this figure was over USD 117 million in SNNP. **Only 3 percent of the total drought-associated loss in the livestock sector comes from the remaining areas in the country** (Figure 6).

Cattle, sheep and goats tend to be among the most highly affected livestock types, while camels, horses and mules and poultry are among the less vulnerable groups (Figure 7). Over 6 million cattle died in each drought year (2008 and 2011), 4.5 million of them from drought-related disease. This amounts to 13 percent of the overall national cattle herd in 2008, and 11 percent of the cattle herd in 2011. Furthermore, over 8 million sheep died in 2008 and another 5 million in 2011, which means the country lost cumulatively 35 percent of its national sheep headcount. Overall, during the protracted drought episode, Ethiopia lost 21 percent of its livestock count in 2008 and a further 16 percent in 2011.

Overall, during the protracted drought episode Ethiopia lost 21 percent of its livestock count in 2008 and a further 16 percent in 2011

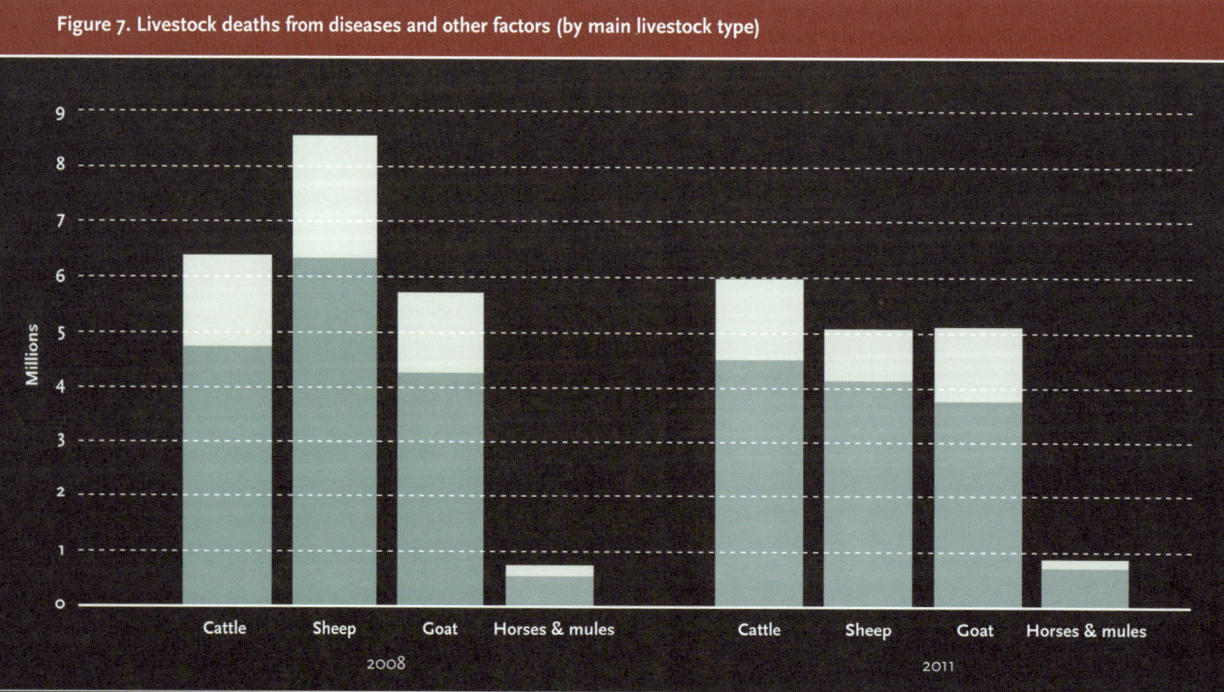

Figure 7. Livestock deaths from diseases and other factors (by main livestock type)

Millions

9
8
7
6
5
4
3
2
1
0

Cattle   Sheep   Goat   Horses & mules   Cattle   Sheep   Goat   Horses & mules

2008   2011

Legend: Deaths from disease, Deaths from other causes

Comparing results for Ethiopia – how do we fare?

Despite placing a significant strain on the crop and livestock sectors, the impacts of the 2008 and 2011 drought episodes in Ethiopia have not been quantified in detail and primary data was not systematically recorded. Government humanitarian requirements reports from 2008 and 2011 testify to the impact of drought in terms of reduced crop and livestock production, which provides a rough basis for comparison. The Government requested about USD 20 million from international donors in order to address the impacts on crop production. Given that such requirements aim to address the most urgent post-disaster needs, it is reasonable to estimate that the full value of crop damage and loss caused by the drought was in fact higher. The FAO estimation of USD 27 million would therefore tend to reflect the value of full sector restoration.

The Government requested about USD 20 million from international donors to address impacts on crop production, and another USD 867 million for livestock health interventions

Furthermore, Government reports from 2008 and 2011 indicate weakened livestock body conditions, low productivity of livestock and high animal mortality in the south and southeastern parts of the country. An estimated 7.8 million heads of livestock in the pastoral and agro-pastoral areas are reported as "affected" by water and pasture shortages and related diseases. The required livestock health interventions are estimated to have cost a total of USD 867 million. This suggests that the FAO methodology provides credible results and constitutes a useful tool for damage and loss assessment even in the context of limited data availability.

The result – strengths, challenges and limitations

Having applied FAO's methodology to assess damage and loss from two polar-opposite disaster settings, it can confidently be concluded that the approach provides important and reliable support to DRR policy and decision making.

Nevertheless, the methodology requires additional fine-tuning, and there are challenges and limitations to be addressed. For example, more accurate results can be obtained by adapting the methodology to assess the cumulative effects of multiple and/or simultaneous hazards, integrating land-use maps and remote sensing technologies as an additional source of information, and by improving the availability of baseline data at the household level. Although agricultural censuses and statistics have improved considerably in recent years, the quality of household survey data can fluctuate from country to country, frequently resulting in the availability of only limited historical information. Additional efforts are therefore needed to improve agricultural data collection and reporting at the global, regional, national and subnational levels. **Standardized damage and loss data collection, monitoring and reporting processes should be established for both medium-to-large scale disasters, as well as for recurrent, smaller-scale events.** The challenge remains to integrate the lesser-represented domains of forestry, fisheries and aquaculture into the analysis. While the framework has been set up, prevailing data gaps hamper further trials.

Additional efforts are needed to improve agricultural data collection and reporting at the global, regional, national and subnational levels

It is important to meet these challenges quickly. The need for a more precise understanding of the impact of disasters and crises on agriculture is urgent, as the ongoing drought in the Horn of Africa further demonstrates.

@FAO/Nepal

Chapter IV
Nepal earthquakes: damage and loss in the agricultural crop sector

This chapter takes stock of the devastating Gorkha earthquakes that struck Nepal in 2015 and explores the impacts on agriculture through a local lens. Analysis of damage and loss for smallholders is geared both towards the big picture (using aggregate data) as well as the household-level. A combination of data sources are used within the framework of FAO's methodology for damage and loss assessment, to evaluate the costs of a rapid-onset natural disaster for the crop sector. The impact on rice – Nepal's primary crop – is explored in detail. Local capacity to collect and manage household-level disaster data is fundamental for effective emergency response, DRR policy and action.

Context

Nepal is highly prone to earthquakes

The magnitude 7.6 and 6.8 earthquakes that struck Gorkha in April and May 2015 affected all economic sectors, including agriculture

In 2015, two consecutive earthquakes struck the high elevations of Nepal's central and western regions, triggering hundreds of aftershocks greater than magnitude 4 (Government of Nepal, 2015a). The first struck the Gorkha district on 25 April 2015 at magnitude 7.6, followed by more than 300 aftershocks of which four were greater than magnitude 6. A magnitude 6.8 quake struck the Sindhulpalchok district on 12 May, killing 9 000 people and injuring another 22 000. Physical destruction of infrastructure affected all economic sectors, causing damage to economic assets – i.e. total or partial destruction of existing physical assets – and economic loss, i.e. changes in economic flows arising from the disaster.

Earthquakes typically trigger subsequent, secondary events and processes that can cause widespread destruction, such as floods, landslides, volcanic eruptions and tsunamis. In Nepal the main secondary geohazards were landslides and the resulting dams. According to Nepal's International Centre for Integrated Mountain Development (ICIMOD), which hosted the landslides monitoring task-force, the earthquakes triggered new landslides in areas previously not affected (Shrestha, Bajracharya & Kargel, 2016).

Agriculture accounts for 34 percent of annual GDP and more than 73 percent of the population is involved in farming, mostly subsistence

In the earthquake- and landslide-struck valleys, as in other parts of central Nepal, more than 73.3 percent of the population is involved in agriculture for either subsistence (64 percent) or commercial farming. Agriculture accounts for 34 percent of annual GDP, with farming systems using a mix of crop and livestock production (Government of Nepal, 2015b). Understanding of the damage and loss incurred in the crop sector is therefore key for disaster risk management and to supporting national resilience policies, planning and action.

Crop production in Nepal

Overall, rice is the main cereal crop country-wide, amounting to 46 percent of the cereal-cultivated area and 55 percent of the production share (Government of Nepal, 2013). However, type and yields of crop production in Nepal vary with elevation. Nepal's plains, the Terai – where the bulk of cereals are produced – is the country's most productive region and was fortunate to remain largely unaffected by the earthquakes. Nepal's hilly and mountainous regions were the hardest hit. These regions are divided into three types: low hills, mid hills and high hills (mountains).

Figure 1 illustrates the relative importance of the main crops for farming households at each elevation, showing information on both the frequency of cultivation and harvest quantities, as reported by households for 2014. The total average household crop production (by type) within the hilly regions is relatively modest, given the average household size of five persons. **Rice production varies by elevation, diminishing in importance as elevation increases.** In the mid hills and mountains, rice production depends heavily on rainfall, fluctuating according to monsoon seasons (Ghimire, Wen-chi & Shrestha, 2015).

Figure 1. Average total crop production, in kg

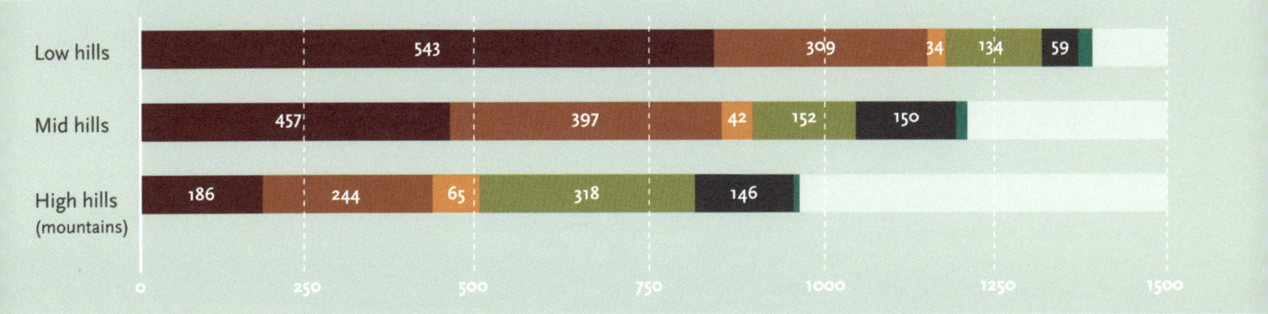

Low hills	543 ... 309 ... 34 ... 134 ... 59	
Mid hills	457 ... 397 ... 42 ... 152 ... 150	
High hills (mountains)	186 ... 244 ... 65 ... 318 ... 146	

Scale: 0 — 250 — 500 — 750 — 1000 — 1250 — 1500

Legend: Rice, Maize, Wheat, Potato, Millet, Pulses, Barley

Figure 2. Crop calendar

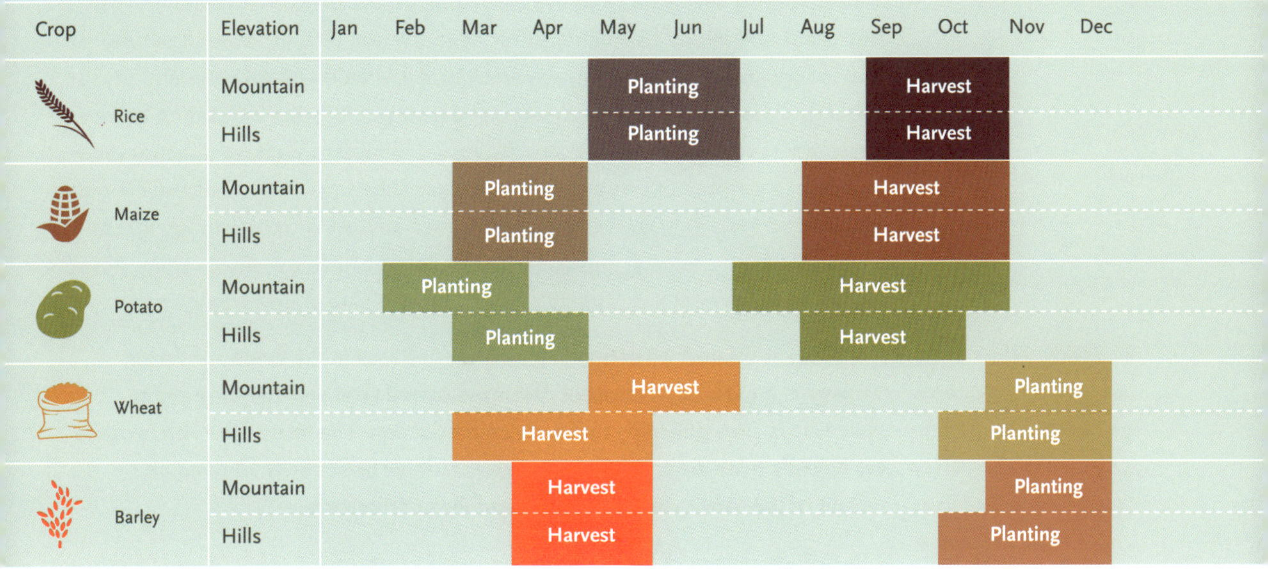

Crop	Elevation	Jan	Feb	Mar	Apr	May	Jun	Jul	Aug	Sep	Oct	Nov	Dec
Rice	Mountain					Planting	Planting			Harvest	Harvest		
Rice	Hills					Planting	Planting			Harvest	Harvest		
Maize	Mountain			Planting	Planting				Harvest	Harvest			
Maize	Hills			Planting	Planting				Harvest	Harvest			
Potato	Mountain		Planting	Planting				Harvest	Harvest				
Potato	Hills			Planting	Planting			Harvest	Harvest				
Wheat	Mountain				Harvest	Harvest						Planting	Planting
Wheat	Hills			Harvest	Harvest						Planting	Planting	
Barley	Mountain				Harvest	Harvest						Planting	Planting
Barley	Hills			Harvest	Harvest						Planting	Planting	

Produced mostly in the mid hills, maize is Nepal's second-largest crop. Potato is the main crop of the high hills. Barley, wheat and pulses represent a minimal proportion of household production across the country, and wheat accounts for only a small share of overall output. The practice of cultivating a mix of cereals, particularly in the higher hills, means that a comprehensive analysis of the disaster on crop production requires assessing the effects of crop substitution as a form of coping strategy.

The earthquakes struck just before the rice-planting season, but after maize and potatoes – the main crops of the high hills– had already been planted.

Nepal's 2015 market and climate

The climate and market situation combined with the timing and location of the earthquakes is especially important in evaluating their impact on agricultural output. To begin with, production is generally lower in the high hills, which were hardest hit by the earthquakes. Second, Nepal's rice production is heavily dependent on its monsoon season (June through September). In 2015, monsoon rains were well below the long-term average (LTA); even May's pre-monsoon rains were less than half the LTA. These drier-than-normal conditions were not abated by August's extremely heavy rainfall (90 percent above the LTA). This combination of drought and earthquake impacts helps explain why paddy-planted area was down by 7.42 percent and paddy production was reduced by nearly 10 percent (Ghimire, Wen-chi & Shrestha, 2015).

Paddy-planted area was down by 7.42 percent, and paddy production was reduced by nearly 10 percent

Third, while the vast majority of Nepal's farming households are subsistence-oriented (only 27 percent report selling crops), and access to markets is generally low in the affected area (16 percent of the affected population is more than two hours away from markets), even this market access was further reduced when the India-Nepal border was closed (September 2015 to January 2016) due to poor relations between the two countries. This may have hindered post-disaster rehabilitation (World Bank, 2016a) by reducing households' ability to both sell output and access essential agricultural inputs such as fertilizer and pesticides (Cosic *et al*, 2016).

Methodology and data

For damage and loss in the crop sector to be calculated one must first know the context-specific availability and use of existing assets (land, irrigation infrastructure, tools, etc.) and inputs (labour, fertilizer, seeds, water, etc.) exposed to natural- or human-induced disasters. Crop productivity is a function of environment, season, labour and management, crop system, and genetics. It is the interaction between these factors that determines economic loss. A disaster may have direct or indirect impact on a number of these factors – such as shifting environment (severe biotic or abiotic stress); management (labour disruption, lack of access to key inputs such as water or chemical inputs); and even the available genetic diversity. Aspects such as management or cropping system may mitigate or exacerbate impacts.

This analysis focuses on rice because it is Nepal's main crop and the least likely to have been newly adopted or substituted for another crop as a result of the earthquakes

This analysis focuses on estimating damage and loss in the crops subsector. Damage is measured in terms of destroyed stored seed and standing crops, while loss is exemplified by reduced rice production. The focus is on rice because it is Nepal's main crop, and the least likely to have been newly adopted as a result of the earthquakes or substituted for another crop.

Externalities caused by the earthquakes, such as increased cost of hired labour, are not accounted for. This means that the results reported here underestimate the true impact of the 2015 earthquakes on Nepal's agricultural production. On the other hand, effects on overall food production and food security may be less severe than expected, because of substitution practices that switch production to relatively less water- and labour-intensive crops, such as maize or millet.

A combination of data sources about agricultural production before and after the natural disasters has been used to estimate damage incurred at the household level, assess causal linkages between that damage and overall production, estimate the monetary cost of economic loss, and control for other important drivers of production such as climate and elevation. The main data source used is the *Nepal Earthquake Response Joint Assessment of Food Security, Livelihoods and Early Recovery* conducted by the Government of Nepal and partners in September –October 2015 (i.e. the October dataset). This dataset includes household production in 2014 and crop expectations for 2015. Data from the

The October and in May datasets are the main sources for this study

Crop Situation Update conducted in May 2015 by MOAD, WFP and FAO is used to estimate the percentage change in aggregate rice production between 2014 and 2015 (i.e. the May dataset).

Furthermore, the analysis employs estimates of the monetary value of damage provided in the PDNA led by the Government in the immediate aftermath of the earthquakes. Nepal's ICIMOD report on the impact of landslides is also used. That report is based on several field surveys, airborne observations and remote sensing mapping to assess the occurrence and impact of the geophysical hazards induced by the earthquakes and aftershocks, especially where they threatened – or had an actual impact on – human settlements and infrastructure. Supporting evidence is drawn from qualitative and quantitative assessments found in the May dataset.

The analysis below is consistent with FAO's methodological framework for damage and loss assessments in agriculture (Annex) and follows the same principles.

Impact of the geophysical hazards on agriculture

83 percent of households reported loss of an agricultural asset. For crops, the immediate impact was destruction of storage facilities, stored seeds and other inputs

The earthquakes caused widespread destruction of infrastructure, houses and animal shelter. As for agriculture-specific damage, 83 percent of households reported damage of an agricultural asset, of which 81 percent reported damage to food and seed storage facilities as well as animal shelters.[1] For crops, the immediate impact was the destruction of storage facilities, along with stored seeds and other inputs such as fertilizer. Over one-third of households lost agricultural tools, posing a potential threat to productivity.

Over a six-month period starting immediately after the earthquakes and ending in October 2015, more than 4 000 landslides were either triggered by the Gorkha earthquakes and their aftershocks or were conditioned by them. The landslides occurred mainly on steep slopes in sparsely populated areas. Only 464 landslides (9 percent) affected human settlements, infrastructure, or irrigation systems. Around 11 percent of the landslides affected croplands, accounting for half of the monetary value of damage caused by post-earthquake landslides. The highly localized nature of landslides warrants continued monitoring and DRR intervention for at-risk communities, which in rare cases can suffer total loss of life and infrastructure.

---

1 Asset damage reported in the survey referred to storage, animal shelters, ploughs, spades, sickles, dokos, livestock carts, irrigation equipment, irrigation infrastructure, aquaculture equipment and infrastructure, mobile telephones, motorbikes, tractors, shop buildings, workshops, generators, carpentry tools, ovens, sewing machines, shelters, household assets or other assets.

Estimating the damage caused by the Nepal earthquake

72 percent of households reporting heavy damage lost rice seed, more than half of them said their stocks were totally destroyed

The quakes' immediate impact on agriculture was damage in terms of destroyed seed and livestock due to the collapse of storage and shelter facilities. There is little doubt that seed loss was caused directly by the earthquakes: 72 percent of households reporting heavy damage in May also reported destruction of stored rice seed, with 51 percent reporting total destruction of their stocks. In contrast, for households sustaining only minor overall damage, relatively few reported seed loss.

Farmers were unable to respond effectively because they lacked the necessary inputs and tools, and had more pressing needs to tend to, such as shelter (FAO & Nepal Food Security Cluster, 2015).

Most of the seed loss was for rice (38 percent) and millet (34 percent). The relative impact of seed loss must be considered in light of both households' typical reliance on stored or purchased seeds, as well as the timing of the disaster.

The seed required to recover pre-earthquake production levels is calculated as follows.

1. The average seed quantities per hectare and average crop yields for each type of crop are drawn from crop-specific studies in Nepal.[2]
2. Average seed loss per household per district comes from the May assessment, taking the average loss for each category to calculate the kg of seed loss (i.e. 0-25 percent = 12.5 percent).
3. These average percentages are incorporated by crop, district, and crop-specific seed requirements as reported in the October dataset, which notes pre-disaster production levels for each crop. This produces an initial estimate of the quantity of seeds necessary to reach pre-disaster production.
4. The volume of lost seed is then aggregated by multiplying the percentage of households declaring seed loss in the October assessment by average household loss, and the number of households in the district. The resulting estimate is considered to be a realistic representation of the damage attributed to the earthquakes, given the high overall reliance of households on stored or exchanged seeds.

Damage caused by lost seed was most severe for rice: 7.8 kg per rice-cultivating household on average

This shows that damage caused by lost seed was most serious for rice, Nepal's main crop: 7.8 kg per rice-cultivating household on average, for a total volume of 1 856 tons of destroyed rice seed. For potato seed, which is heavier and bulkier, damage was 11.4 kg per household on average, for a total volume of 1 869 tons of destroyed potato seed. Notably, potato is the main crop in the higher-elevation districts, which were those most severely affected by the quakes.

Having estimated the volume of destroyed stored seeds, a combined cost of economic damage in the crops subsector can now be derived. Using the average cost of food grains by kg to estimate the value of seed loss, it is determined that the total cost amounted to USD 2.3 million. This does not include the transportation cost of seed distribution, which may be high in the severely affected mountains where households are remote from markets.

---

2 See bibliography for extensive list of crop-specific sources, except barely, which was calculated based on seed requirement yields available for wheat from other sources.

Figure 3. Seed loss (% of households reporting loss)

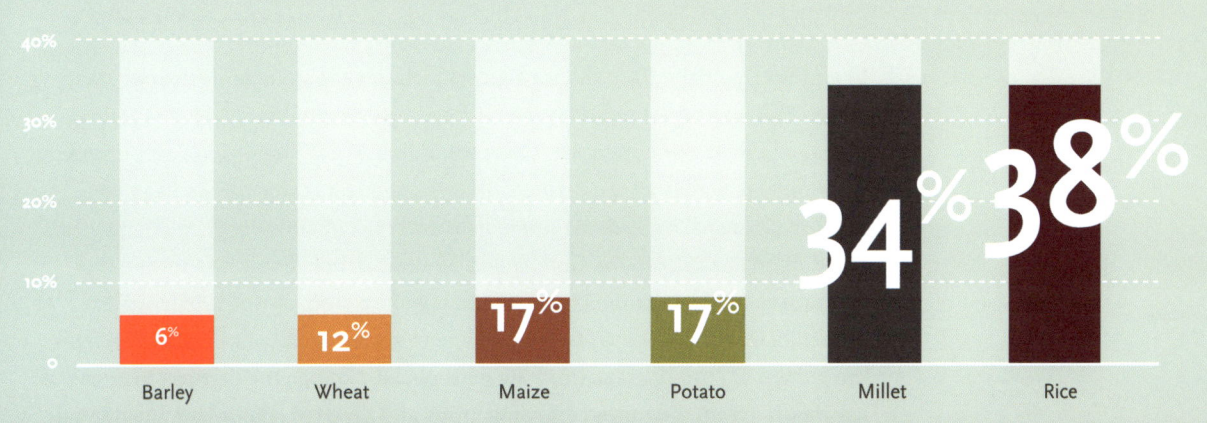

	Barley	Wheat	Maize	Potato	Millet	Rice
	6%	12%	17%	17%	34%	38%

Table 1. Total damage – value of destroyed seed by crop

Crop	Volume of destroyed seed (kg)	Total damage (kg)	Total economic cost (USD)
Rice	7.8	1 856 932	835 619
Maize		1 030 015	309 005
Millet	3.2	662 996	205 529
Wheat	8.9	871 447	409 580
Barley	3.9	15 484	4 335
Potato	11.4	1 869 406	560 822
Total		6 306 280	2 324 890

Figure 4. Standing crop loss (% of households reporting loss)

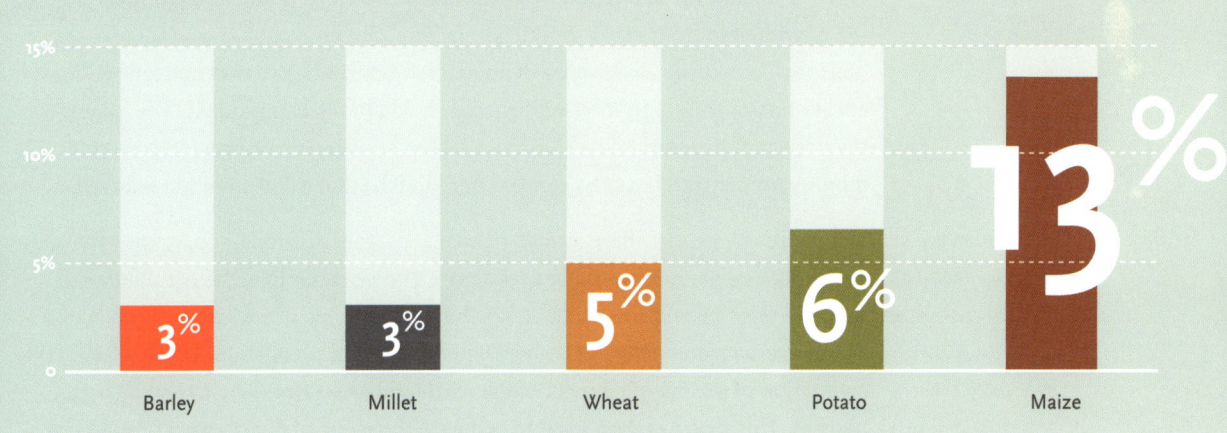

	Barley	Millet	Wheat	Potato	Maize
	3%	3%	5%	6%	13%

This points to a related issue: the high cost of transporting potato and other seeds to mountain areas. Given the inaccessibility of the remote high elevations, it is particularly important to have a functional network of local input suppliers who can pool transportation costs, thereby reducing replacement costs for farmers.

In addition, damage from the earthquake includes the value of destroyed standing crops. This was minor overall, ranging from 3 percent of households reporting damage to maize or potato, to 13 percent of households reporting some standing crop damage across all the districts. In this context it is worthwhile flagging another biological hazard: the immediate outbreak of armyworms that attacked standing maize in both the Gorkha and Sindhupalchok districts following the earthquakes. While the direct damage to standing crops is not substantial in monetary terms, compared to the damage to stored seed volumes, it completes the picture of crop damage from the earthquake. **Overall, the damage in terms of key inputs (seed), destroyed standing crops and diverted labour had substantial and long-lasting effects on harvests up to six months after the disaster had struck.**

Estimating the crop production loss caused by the Nepal earthquakes (focus on rice)

This section attempts to establish – as a commodity-specific example – the economic loss from reduced rice production caused by geophysical hazards over the entire summer growing season. In order to do so however, it is necessary to disentangle the effect of the earthquakes from other important determinants of agricultural production. The geophysical hazards struck in the poorer high and mid hills, where rice production is typically less than in lower elevations. Furthermore, 2015 offered generally poorer growing conditions due to a delayed and weak monsoon season as well as weakened market access due to the India-Nepal border closing.

The earthquakes made what was already a poor season for most farmers even worse for those in the mid and high hills

When surveyed in October, households in the 11 affected districts expressed low expectations for the upcoming production of rice as well as other cereals. This is in line with farmers' generally below-normal production expectations across all crops and districts, that season – no surprise given Nepal's market and climate conditions that year. However, households reporting damaged assets (indicating impact from the earthquakes) tended to have more negative expectations than those reporting no damage. This shows that the earthquakes made what was already a poor season for most farmers even worse for those in the mid and high hills.

Using a **"difference in differences"** estimation model to separate the impact of the earthquakes from other production factors, the production trend for affected households is compared to that of unaffected households, using household fixed effects (i.e. only the variation within households in the two groups). This way, the impact of the earthquakes on the affected households (as measured with a dummy for damaged assets) can be compared to those that were not affected.

There was an overall average reduction of 50.85 kg of rice produced per household between 2014 and 2015

The results show that there was an overall average reduction of 50.85 kgs of rice produced per household between 2014 and 2015. Of this, the net difference between affected and unaffected households was 17 kg, which can be established as the overall global effect of the earthquakes on rice production. **On this basis, the estimation of the overall impact of the earthquakes on rice is -17.01/-50.85 = 33.5 percent. This figure is then used to estimate the value of direct impact caused by the earthquakes alone on the overall production of rice (Figure 5).**

Figure 5. Rice production loss

Production in 2015 (mt)
290 042 —

Production in 2014 (mt)
355 519 =

−65 477 = 33.5% = 21 935
Aggregate change   Share of change   Aggregate loss
caused by disasters   caused by disasters

45 → { 987 075 000
Price/kg   Monetary value of loss (Rupees)
May 2015
(Rupees)   9.9 million
monetary value of loss (USD)

The economic cost of the earthquakes on lost production in rice alone amounted to approximately USD 10 million. This shows the importance of an efficient agricultural response: the extent of rice production loss (USD 9.9 million) is almost 12 times greater than the direct damage in terms of seed destruction (USD 835 619), despite an efficient response that largely mitigated economic loss. The effect on household incomes was much larger, and this is far more significant.

The positive impact of the rapid aid response must be mentioned. Over 82 percent of households reported having received assistance such as rapid rice seed distribution alongside other food and non-food aid packages. Given that distributed rice seed tends to be of high quality and drought-resistant, it can generally be assumed to produce higher yields than seed used before the quakes.

Households that received assistance produced on average 48 kg more rice than those that did not. Rarely is such a high impact of seed distribution seen on harvests

Having zoomed in on the effect of agricultural inputs, namely seeds, the impact of assistance received is apparent: affected households that received assistance produced on average 48 kg more rice than those that did not. Rarely is such a high impact of seed distribution seen on harvests. Research shows that risk-averse farmers are generally reluctant to adopt new rice varieties, and aid distributions are usually not the only alternative source of seed. In the case of the Gorkha earthquakes, however, the large loss of rice seed and the little time left for planting likely promoted quick adoption. This means that while assistance was properly targeted and highly efficient in mitigating loss, it may be that the positive impact of the assistance cannot be generalized to other disasters where farmers had a choice of seed varieties and access channels.

It is important to note that aggregate figures are heavily skewed towards large producers. **Even if the severity of aggregate loss may be a close approximation of the severity of household loss, the impact of such loss on household production touches human lives more directly than the impact on aggregate production.** Indeed, the regions most affected were populated with relatively small landowners and subsistence farmers with little access to markets, for whom an average loss of a quarter of their main staple is very important. These regions also typically had a smaller amount of crop production overall, and therefore a high risk of food insecurity.

Overall, the "difference in differences" model confirms that the earthquakes had a negative impact on rice production, distinct from other influencing factors (poor rainfall, trade disruption). Because the affected areas were not large rice producers, the impact is modest on a per-household average. However, loss was also accompanied by damage to seed storage and irrigation systems. Finally, these findings do not take into account the externalities caused by the disasters (for example potentially higher wages, transport disruptions caused by road destruction, and other effects that impact households only indirectly). Therefore, the results almost certainly underestimate the true impact of the earthquakes.

In sum, despite limited impact on the agricultural sector compared to the overall impact, the Gorkha earthquakes did cause significant loss of crop production, as well as substantial damage in terms of stored crops and standing crops.

These can be mainly attributed to damaged seed storage units, the inability to substitute household stocks with market supply, and the diversion of labour away from subsistence agriculture. The latter accentuates a worrying trend already visible in the Nepali agricultural landscape.

Estimating the monetary value of agricultural loss

On average across the 11 districts, 33.5 percent of household loss was due to the earthquakes

A further, and more relevant, estimate of economic loss can be calculated based on household loss, rather than aggregate loss. Using the same prices as above, the value of the change in rice production can be estimated and compared to monthly household income as indicated in the October *Joint Recovery Assessment* dataset. **The average loss for households affected by the earthquakes was one month of household income, ranging from a low of 23 percent of monthly income in Okhaldhunga, the least-affected district, to a high of 227 percent of monthly income in Gorkha, the epicenter of the most powerful quake.** This is despite the rapid response provided by humanitarian agencies for both food and seed distribution, as described in the next section. As already noted, on average across the 11 districts 33.5 percent of household loss was due to the earthquakes (although there were obviously large variations among districts, with households in the most affected districts having incurred major economic loss due to the earthquakes).

Table 2. Household loss by district

District	Severity of loss	Mean loss in kg	Mean household economic loss (USD)	% of monthly income
Dhading	-20%	-95.22	-4 285.04	-52%
Dolakha	-10%	-38.31	-1 793.25	-29%
Gorkha	-58%	-300.62	-13 370.90	-227%
Kavre	-12%	-83.18	-3 834.03	-52%
Makwanpur	-19%	-110.59	-4 262.61	-91%
Nuwakot	-21%	-132.77	-6 006.67	-108%
Okhaldhunga	-17%	-32.83	-1 485.28	-23%
Ramechhap	-28%	-108.87	-4 926.67	-118%
Rasuwa	-33%	-164.85	-7 163.12	-103%
Sindhuli	-33%	-178.66	-8 369.28	-161%
Sindhupalchok	-32%	-163.44	-6 879.73	-106%
Overall average	-26%	-137.50	-6 090.93	-103%

The lack of household data on 2015 production impedes further analysis on the impact of the earthquakes by household characteristics and resilience, but it is clear that households in the worst affected districts suffered heavy loss.

Although lacking the necessary data to assess the impact of the earthquakes on other crops, the October dataset offers a key indicator. It shows that households in the 11 districts surveyed had overwhelmingly negative crop expectations for 2015, which is true for all crops for which data are available. **This suggests – strongly – that households producing those crops likely incurred significant loss.** This would have had dramatic consequences on food security and poverty for households without humanitarian support. There is anecdotal evidence, for example, that maize was left to rot because no labour was available to harvest it in the immediate aftermath of the quakes.

Nepal earthquake response in the agricultural sector

Agricultural assistance reached 240 000 households in the six most severely affected districts

The response in the agricultural sector was a successful one. Immediately after the earthquakes, FAO actively supported Government-led efforts to address and coordinate the response to food security and livelihood needs in the six most severely affected districts: Sindhupalchowk, Nuwakot, Dhading, Gorkha, Rasuwa and Dolakha. Between April 2015 and January 2016 alone, 240 000 households in these districts were provided with agricultural assistance, including the immediate distribution of rice seed to meet the imminent planting window, as well as later seed distributions for other crops. A generally good state of preparedness of the Government and humanitarian agencies allowed for rapid assessments and a swift response. The Government of Nepal had carried out disaster-preparedness simulations (for floods) and had already set up emergency coordination cells at the district level. Furthermore, the response strategy was well-coordinated between the Government and key humanitarian agencies, while local partners had a central role in decentralized implementation.

Substantial crop loss threatens food security, requiring increased monitoring and rapid response

This case study demonstrates that **the economic loss caused by the earthquakes was substantial. The impact on household production appeared much more severe in relative terms than the overall USD impacts on the crop sector, since it affected subsistence farming, livelihoods, and food security.** The reason is that aggregated production is highly skewed towards less vulnerable, large-scale farmers. Greater monitoring is necessary at the household level to better understand the impact of disasters on crop production and the interaction between these effects and other productive factors.

Monitoring the impact of the disaster on crops

→ Household-level data are fundamental in order to better understand the impact of disasters on agricultural livelihoods and identify the most at-risk categories of farmers. Aggregate figures – which are heavily skewed towards the big picture – do not provide a reasonable estimate of loss for smallholders, who constitute the vast majority of the population in Nepal and many other countries, often the poorest.

→ Integrating agricultural modules into ad-hoc multi-sectoral surveys is a cost-efficient option that has proved effective in Nepal. Working with the Central Bureau of Statistics to integrate a food security module in annual household surveys is another option that should be explored.

→ For ad-hoc post-disaster assessments, including baseline indicators – as for the October *Joint-Recovery Assessment* – allows for a rigorous methodology through the "difference in differences" approach. This good practice should be promoted.

→ In addition to quantitative surveys, the qualitative assessments provided by the emergency response team in May 2015 were valuable in enriching the study. This good practice should be promoted and used to highlight information needs for quantitative data collection.

Lessons learned on the economic, social and environmental impact of the disaster

→ The immediate direct damage to crops by the Gorkha earthquakes was limited – mostly seeds and very few standing crops – yet the actual economic loss over the whole season amounts to over 12 times as much as the direct damage, despite an efficient response that largely limited loss. This underlines the importance of rapid response in agriculture to prevent total crop failure.

→ The direct damage to the agricultural sector in terms of seed loss and irrigation had a substantial impact on crop production by reducing the amount of physical inputs. This was further amplified by the shortage of labour available for farming.

→ Responses to earthquakes should therefore include the provision of labour-saving technologies as well as access to inputs. Farmers should be given a choice of seed varieties that suit their needs and preferences; voucher systems with a network of input suppliers may be a good option.

Lessons learned on preparedness for disaster reconstruction

→ Disaster preparedness in Nepal enabled a swift response: trained field assessment teams in all districts were able to conduct rapid assessments that helped determine needs and target beneficiaries within days of the disaster. The presence of Government district emergency officers supported the evaluation of damage and the coordination of aid. Both are good local practices that should be promoted.

→ Landslide monitoring must be set up immediately after earthquakes, and prevention measures implemented immediately in zones already identified as at-risk for landslides, as well as those newly placed at risk by the earthquake.

India 2005 〈 India post-tsunami

Chapter V
The impact of disasters on fisheries and aquaculture

This chapter focuses on integrating fisheries and aquaculture into FAO's disaster impact assessment methodology, as part of a holistic approach to the sector. Of the 23 percent of disaster impact absorbed by agriculture, 6 percent is incurred by fisheries and aquaculture. Yet there is no systematic approach to monitor damage and loss in the subsector and data are seldom collected. This is a key challenge to securing the subsector's place on both the global and national DRR agendas.

Impact of disasters on the fisheries and aquaculture sectors

Impacts on productive assets

Fishing communities, ports, harbours and aquaculture installations are commonly situated at the interface between water bodies and land, precisely where various hydrological and meteorological disasters strike. Tsunamis, tropical cyclones, storm surges, coastal and general floods have the greatest impact on the sector. **Human-induced disasters such as toxic and chemical spills, nuclear plant accidents, land-based pollutants, and wilful discharge of oil, petroleum products and chemicals also impact fisheries heavily, and may affect the health of the entire aquatic ecosystem**. Exposure to such hazards makes fisheries highly vulnerable. Meanwhile, the necessary post-disaster rehabilitation is an expensive and complex process, especially if it involves clean-up of toxic waste.

The scale and complexity of fisheries operations carried out within a given area is conditional on the level of local technological and socioeconomic development, the surrounding environment, the health of the ecosystem, and the availability of fisheries resources. Therefore, **the impact of natural disasters on fisheries and aquaculture is greater in vulnerable areas where poverty is prevalent, infrastructure poor, population density high and adequate DRR strategies lacking.** In general, all types of natural disasters can have varying effects on fisheries and aquaculture depending on the exposure and intensity of the hazard.

Impacts on ecosystems and on fisheries/aquaculture living resources

Natural disasters also affect marine and inland water ecosystems. Some ecosystems can suffer significant damage, as was the case for the March 2005 earthquake in Indonesia that ravaged and uprooted coral reefs. Coral reefs can be severely damaged by the force of the waves themselves and by the heavy debris and boulders that land on them destroying corals (McAdoo *et al*, 2011).

Drought in distant inland areas can also negatively affect the quantity and quality of coastal waters when the outflow of rivers is greatly reduced, diminishing the quantity of organic matter that would normally flow to the coasts and thereby decreasing primary productivity around river mouths.

Meanwhile, floods can have both negative and positive effects on fish populations. Major flooding can overwhelm water management systems such as dams and reservoirs, thus causing transfers of land-based pollutants to coastal and inland water bodies, with harmful effects on aquatic animals and the ecosystem. **Floods also wash away fish eggs and larvae. Fish may be damaged when coming into contact with obstacles in the floodwaters and excessive silt temporarily impedes fish from finding food.** On the other hand, the increased amount of water during floods can provide more volume and areas for hiding, foraging and growth of fish populations.

Table 1. Types of damage to fisheries and aquaculture showing damage to productive assets and primary and secondary impacts on ecosystems

Disaster type	Damage to upstream and downstream productive assets	Primary and secondary impacts on aquatic ecosystem
Cyclone and storm surge	Damage and destruction of coastal and inland fishing and fish transport boats, engines and fishing gears. Damage to aquaculture structures such as ponds, cages and shellfish and seaweed growing systems. Damage and destruction of buildings and infrastructure (these include harbours, jetties, onshore processing plants, drying racks, smoking houses, ice factories, boat sheds, electrical supply, fishery supply stores, fuel storage and pumping, cold storages, refrigeration equipment, fish transport vehicles and others).	Ghost fishing, i.e. when lost or unattended gear such as nets, long lines, and traps continue to catch fish and deplete stocks without anyone benefiting from the catch. Loss of farmed aquatic plants and animals. Damage to beaches and nesting areas, sand dunes, coastal shrubs and trees.
Coastal flood and inland flood	Damage and destruction of coastal and inland fishing and fish transport boats, engines and fishing gear. Washing away of coastal and inland aquaculture ponds, cages and equipment (aerators, generators, etc.). Damage and destruction of hatcheries and feed stores. Loss of fish that are in the ponds and cages. Damage and destruction of buildings and infrastructure (see above).	Increased amount of water provides more volume and areas for hiding, foraging and growth. Coming into contact with obstacles in fast-moving floodwaters may damage fish. Floods bring land-based pollutants (plastics, garbage, pesticides, chemicals, debris, fishing gear, etc.) onto coral reefs, coastal waters and inland lakes, rivers and reservoirs, causing damage and possible ghost fishing.
Tsunami	Damage and destruction of coastal fishing and fish transport boats, engines and fishing gear and aquaculture facilities (ponds, cages, shellfish rafts and longlines). Damage and destruction of buildings and infrastructure (see above).	Boulders thrown onto coral reefs cause damage and destruction. Waters receding back into the sea after the tsunami transport land-based pollutants (plastics, garbage, pesticides, chemicals, debris, fishing gear, fuel, etc.) onto coral reefs and coastal waters, causing additional damage to reefs.
Earthquake	Damage and destruction of fish processing plants, aquaculture laboratories, hatcheries, ice plants.	Earthquakes cause coral reefs to be uprooted.
Drought	Reduced water quantity and quality leading to reduced production in aquaculture and inland fisheries.	Organic materials transported by rivers are diminished because reduced water flow causes low primary productivity in coastal areas. This can impact fisheries in other areas. Reduced water in rivers, lakes and reservoirs concentrate fish in smaller volumes of water making them easier to catch; reduced amount of food for growth.
Harmful algal blooms	Kill-off of important fisheries targets fish and shellfish species as well as farmed shellfish and fish.	The effect of the algal bloom depends on the species (especially those containing toxins), the geographical extent and duration of the bloom, as well as the strength and direction of the prevailing currents at the time.
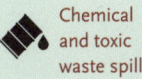 Chemical and toxic waste spills	Contamination of shellfish beds, downstream kill-off of aquaculture fish and shellfish stocks and damage to fishing gear. Risk of fire from petroleum and toxic fumes.	Kill-off of fauna in mangrove swamps. Contamination of beaches, nesting areas, birds. Kill-off of highly valuable commercial species.

An analysis of 74 large-scale post-disaster assessments between 2006 and 2016 conducted by FAO found that the fisheries and aquaculture sector is highly vulnerable to disasters. Of these, **storms (such as hurricanes), floods and tsunamis caused roughly 82 percent of the overall damage and loss to the fisheries subsector.** This amounted to over USD 1.1 billion, which represents around 3 percent of all damage and loss within the agriculture sector. Furthermore, **over the last 20 years, disease outbreaks have reportedly cost the aquaculture industry tens of billions of dollars.**

Storms, floods and tsunamis caused roughly 82 percent of the overall damage and loss to the fisheries subsector

Measuring disaster impact on fisheries and aquaculture

In line with the increasing importance attached to DRR and DRM as cornerstones of development and emergency work, the fisheries and aquaculture sectors must be better integrated into national resilience policy as well as into the measurement of disaster-related targets of international initiatives such as the Sendai Framework and the SDGs. In addition, it is important that rehabilitation and recovery be undertaken within the context of sustainable fisheries and aquaculture resources management and that they not be divorced from fisheries policy and planning. Rehabilitation and recovery should be at a level that does not jeopardize the medium- and long-term livelihoods of the victims of the disaster. *The Code of Conduct for Responsible Fisheries* (CCRF) and the FAO *Voluntary Guidelines for Securing Sustainable Small-Scale Fisheries in the Context of Food Security and Poverty Eradication* provide the groundwork and overall framework for both climate change adaption and building resilience against natural and human-induced disasters in small-scale fishing communities.

An exhaustive overview of damage and economic loss from disasters in the fisheries-aquaculture sector can only be calculated with sound pre- and post-disaster data. However, more often than not, such data are either lacking or incomplete. In an effort to better understand how to integrate fisheries and aquaculture in damage and loss assessments, this chapter presents a review of existing methodologies. It also lays the groundwork for integrating a fisheries/aquaculture component into FAO's damage and loss assessment methodology (Annex), and thereby for measuring progress towards the internationally agreed targets of the Sendai Framework (indicator C-2) and the SDGs (target 1.5.2).

How far do existing methodologies go?

While the PDNA is recognized as the methodology of choice and constitutes a useful tool for immediate assessment of post-disaster needs, actual coverage of the impact borne by the fisheries and aquaculture subsector is limited. What is more, PDNAs are only carried out for large-scale disasters, if data are available.

There are several shortcomings at present: a lack of fisheries expertise in the teams conducting the PDNAs; a lack of baseline data for calculating the effects; and unreliable or incomplete data leading to inaccurate estimates. Other data sources on economic loss in the fisheries and aquaculture sector are extremely limited. While there are a number of general databases on disaster loss (such as CRED; the MunichRe and SwissRe databases; and the Global Facility for Disaster Reduction and Recovery, GFDRR), none of these systematically collect disaggregated data on fisheries and aquaculture. **There is currently no systematic approach for the continued monitoring of damage and loss in fisheries and aquaculture.**

Quantifying damage and loss

While it is clear that the fisheries and aquaculture sector is particularly vulnerable to disasters, a solid system for the assessment and quantification of post-disaster damage and loss has not yet been established. **FAO is addressing this gap and providing a more holistic representation of the impact of disasters on fisheries and aquaculture by developing and applying its damage and loss assessment methodology (Annex).**

In addition to the direct damage experienced by the sector, this methodology also addresses other economic post-disaster impacts on fisheries and aquaculture – such as loss of markets, lack of fuel, and either increased or depressed selling prices for fishery products – that can severely impact food security locally or even at the national or regional levels. The extent of the impact is dependent on the scale of the disaster, the size of the country and whether other regions in the country can increase production to make up for the loss caused by the disaster. Calculating direct and indirect linkages between natural disasters and production loss is only possible with a full understanding of the economic parameters of markets, fishing operations, value adding, fishing seasonality, demand and supply, fisheries management measures such as spatial and temporal closures, amounts of available and unaffected products in cold storages and the health of fish stocks in general.

The extent of the impact depends on the scale of the disaster, the size of the country and on whether other regions in the country can increase production to make up for the loss caused by the disaster

Information requirements for improving damage and loss assessment in fisheries and aquaculture

Identifying the particular data and information needs is the crucial first step towards a sector-specific approach to assessing the impact of disasters in fisheries and aquaculture. Table 2, adapted from FAO's 2013 *Guidelines for the Fisheries and Aquaculture Sector on Damage and Needs Assessments in Emergencies*, presents the fundamental data elements necessary to devise a sector-specific methodological approach in line with the requirements of the PDNA methodology. **This information should be collected by governments and updated on a yearly basis. The more precise the data, the higher the quality of the sector assessment will be, allowing for an efficient rehabilitation and recovery process.**

Productive asset	Pre-disaster information requirements
Boats	→ Number of boats by type, length and age of the fishing vessels, by fleet segment.
	→ Number and type of government fishery research vessels, coast guard and navy patrol vessels and value of each vessel.
	→ Type of construction material of the vessels and cost of construction, including material and labour of fishing vessels and government vessels.
	→ Typical engine make and horsepower by boat type and fleet, including cost of engine and installation.
	→ Location of boats by number and fleet.
	→ Design of vessels, including drawings and photos.
	→ Fish storage capacity of the vessels and cost of on-board refrigeration equipment.
	→ Imported price if the vessels are not built locally.
	→ Insured value of boats by type and area.
Boat equipment	→ Description and cost of equipment needed for fishing operation.
	→ Electronics including radio, GPS, sounder, radar.
	→ Safety equipment (life jackets, life rafts, distress signals and first aid).
Fishing gear	→ Inventory of numbers and types of fishing gear used by each fishery.
	→ Number and type of ancillary gear needed for each operation.
	→ Cost of unit of gear by type.
	→ Number, type and depth of Fish Aggregating Devices.
Fishing operation	→ Fisheries costs and earnings evaluation studies per fishery type and fleet segment; this includes costs (investment, maintenance, fuel, deprecation, repairs, licensing, transport, etc.) and earnings (catch-rates species, operational days, vessel sales, etc.).
Fisheries management	→ Fisheries production trends by fleet segment and species.
	→ Status of fisheries resources and fisheries management measures, including community-based approaches and co-management.
	→ Catch per unit effort, by gear, per vessel and fleet segment.
	→ Valuation of the maximum sustainable yield (MSY) of the fisheries resources in terms of biomass and value of that biomass in financial terms.
	→ Projections of future production based on trends and fleet structure and growth. Biological and resource assessments should also be valued in monetary terms.
Fishing gear stores	→ Number of stores.
	→ Value of monthly inventory.
	→ Annual turnover/sales.
	→ Insured value of the store and stocks.
Boat-building	→ Number of boat-building yards by type and number of boats built annually, location of the yard and the final value by type of boat.
	→ Number of boat-builders working per boat year and annual turnover.
	→ Insured value of the yard, equipment and materials.
	→ Value of boat sheds, tools, wood and equipment normally in stock at the boatyard.
	→ Number of slipways, dry docks, the age and value of each installation.
	→ Number of haul-outs and repairs done annually by boatyard.

From assessment to action – building resilience

It is not enough to calculate economic damage and loss and then use these figures directly for rehabilitation and recovery within the sector. Fishery resources are often heavily exploited and the status and trends in certain fisheries are not well known. The post-disaster rehabilitation and recovery of a fishery should be conducted with a precautionary approach, since replacing damaged and lost items with new ones actually increases the efficiency of fishing operations, thereby increasing the detriment to an already heavily-exploited resource.

Therefore, it is necessary that the information requirements in Table 2 be complemented by supporting information, so that policy decisions can ensure a sustainable level of rehabilitation and recovery in line with the CCRF and Voluntary Guidelines. For example, it is important to take stock of: previous disaster preparedness work and the livelihood profiles of affected communities; population information and demographics; the current institutional framework; and fisheries and aquaculture policy and management. Furthermore, **it is important to incorporate existing fisheries baseline surveys or censuses as well as sector value chain studies and analyses.** These are an integral part of damage and loss assessment, since they show the economic value of fish and fisheries products and identify key value chain stakeholders.

Overall, it is important that a **solid information system and data-based analyses guide a streamlined damage and loss assessment in fisheries and aquaculture, leading to sustainable technical solutions to disaster impact.** This will contribute to building capacity for reconstruction and recovery as well as strengthening the overall resilience of the sector.

Applying the new methodology (Annex) will allow for better informed national resilience policy and action, addressing damage and loss as well as considering related recovery and rehabilitation costs. It will also contribute to monitoring the respective international DRR targets under the Sendai Framework and the SDGs.

Assessing fisheries damage and loss from Cyclone Evan in Samoa

Tropical Cyclone Evan passed over Samoa as a Category 2 tropical cyclone on 13–14 December 2012, damaging or destroying nearly 1 700 houses and affecting more than 8 000 people (OCHA, 3 Jan. 2013). The cyclone's intensity increased as the storm went on to strike Wallis and Futuna Islands and Fiji. Across all sectors the total damage was calculated at USD 103.3 million and the total loss at USD 100.6 million (currency rate as of Dec. 2012). The fisheries and aquaculture subsector accounts for around 3.4 percent of the disaster's overall damage and loss.

Agriculture contributes about 10 percent of Samoan GDP, but it employs around two-thirds of the national labour force and is an important source of household income. According to pre-disaster surveys, 24.8 percent of households were engaged in fishing, with most of the catch used for household consumption. The inshore coastal fishery is important for village economies. An estimated 9 557 metric tonnes valued at USD 32.85 million are caught by subsistence fisheries. Few households are engaged in commercial fishing: in 2011, commercial catches were estimated at 2 402 tonnes, valued at USD 7 million.

About 75 percent of the agricultural area in Upolu was either severely or moderately affected by Cyclone Evan, and the crop subsector was hardest hit both in terms of damage and of loss. Flash flooding in localized areas wiped out or heavily damaged a number of farms. The total impact calculated for the agriculture sector (including crops, livestock and fisheries) amounted to USD 30.5 million (USD 4.2 million in damage and USD 26.2 million in loss). Just under 10 percent of this was in fisheries, where total damage and loss amounted to USD 3 million (USD 850 000 in damage and USD 2.2 million in loss). The damage and loss in fisheries is estimated at about 30 percent of total agriculture sector GDP for 2013 (crops, livestock, and fisheries).

Damage to fisheries has largely been in the artisanal sector, with about 27 percent of canoes owned by artisanal fishers reported as damaged. The fishers also reported that about 50 percent of their fishing gear was destroyed. Most of the damage to canoes was caused by fallen trees, while fishing nets (usually stored outdoors) were damaged by debris. In the commercial fishing sector, 12 out of 63 boats were damaged – either by fallen trees or from being rammed against seawalls during the cyclone. Freshwater aquaculture farms also sustained damage: twelve of the 51 existing tilapia ponds were damaged, mainly due to flooding.

As far as loss is concerned, the largest fisheries production loss was accrued within artisanal fishing activities, where fishers are estimated to have lost income for one to six months after the cyclone, the time it took for their canoes and gear to be replaced.

Despite the relatively lower cost of damage in fisheries (at USD 850 000), repercussions for the sector's livelihoods are far-reaching. Imputed income decline analyses for wage-salaried and self-employed persons show that livelihoods in fisheries were particularly impacted, incurring an overall 49 percent decline in 2013 before returning to roughly 92 percent of pre-cyclone levels in 2014.

A targeted donor-led PDNA was undertaken. Comprised of technical experts in agriculture, including fisheries, the PDNA team provided tangible recommendations for fisheries rehabilitation.

→ Compensation for loss of canoes and fishing gear. This was made to households directly or through a voucher system operated by village fisheries management committees. It is important to note that overfishing of herbivorous fish by subsistence fishers, damage to corals by the 2009 tsunami and the crown of thorns starfish were already taking their toll on the fisheries. **The rehabilitation therefore focused on providing limited equipment and then only to fisher communities that were participating in community-based fishery management programmes.** The communities that did not participate in those programmes received only agriculture inputs. In this way the rehabilitation did not increase fishing on stocks that were already vulnerable to over-exploitation.

→ Transfer through the matching grant programme. Aquaculture farmers who lost their ponds and assets could apply for up to 70 percent of their loss up to a maximum of USD 7 000 (or such percentage / amount as agreed by the government and The World Bank).

→ A disaster vulnerability reduction support programme was formulated, to be implemented directly by the fisheries.

Chapter VI
The impact of disasters on forest resources

Forestry is another subsector on the fringes of post-disaster damage and loss assessment. Significant knowledge and data gaps hamper systematic reporting of disaster impact to forests, and existing PDNA methodologies largely bypass the subsector. This chapter pieces together key knowledge, facts and figures in an attempt to bring forestry into the disaster impact assessment discussion. It consolidates relevant subsectoral perspectives, identifies major knowledge gaps and proposes steps towards improving the assessment of forest-related damage and loss.

What we know about impacts of disasters to forests

There are about four billion hectares of forests in the world (FAO, 2015a), and millions of them are affected by some form of natural disaster or disturbance every year. While the universally adopted definition of disaster pertains to. "a serious disruption of the functioning of a community or a society at any scale, due to hazardous events interacting with conditions of exposure, vulnerability and capacity and leading to human, material, economic and/or economic losses and impacts"(UNISDR, 2017), in the context of forestry, certain adjustments may be required.

In order to understand how the sector is impacted by natural hazards and disaster events, it is important to understand what constitutes a disaster in a forest setting. The term "disturbances" refers to a range of detrimental impacts in the sector. Forest disturbances encompass both the environmental fluctuations and the destructive events that disturb forest health and/or structure, and that trigger a change in resources or physical environment at any spatial or temporal scale (FAO, 2004). Under normal circumstances, disturbances can be an integral part of the forest ecosystem. However, catastrophic disturbances can have undesired impacts on forest ecosystems and affect environmental functions, with consequences for biodiversity as well as livelihoods and climate change impacts. It is these catastrophic disturbances, which fall in line with the universal notion of disaster, and their devastating impacts that are at the centre of FAO's analysis of damage and loss in the forestry sector.

According to recent FAO estimates, fire devastates 70 million hectares of forest every year, mainly in tropical America and Africa.

While aiming to lay the foundations for evaluating the impacts of disasters on forestry and ultimately establish a systematic methodology for forestry damage and loss assessment, it is important to take stock of those phenomena that have disastrous impacts on forests. According to recent FAO estimates (FAO, 2015a), fire devastates 70 million hectares of forest every year, mainly in tropical America and Africa. Severe weather affects almost 40 million hectares, while insect pests can ravage up to 85 million hectares, mainly in temperate North America. Diseases, on the other hand, affect 12.5 million hectares per year, mainly across Asia and Europe.

Types of disasters affecting forest systems

There are five major abiotic and biotic disturbance factors that cause damage and loss in forestry, and they broadly correspond with the standard natural disaster categories (see Introduction), including anthropogenic factors and human-induced disasters:

→ Meteorological events (cyclones, tornadoes, wind-, snow-, hail-, dust-, sand- and thunderstorms, etc.) Storms, cyclones and hurricanes are among the most common disasters in this category, and single events can cause considerable loss. In Europe, storms account for more than 50 percent of all damage to forests and the problems are often aggravated by further management of forests and cultivated areas. In China, storms pose frequent problems to forests and even a single event can have far-reaching consequences, as the 2008 ice storm proved, destroying 20 million hectares of forest (10 percent of national forest cover), and 373 000 hectares of winter crops (roughly 40 percent) (Millennium

Ecosystem Assessment, 2005). Other striking examples include hurricanes Katrina (Louisiana 2005) and Matthew (Haiti 2016). Thunderstorms and lightning often bring heavy precipitation and rain and may also trigger tornadoes, while lightning can cause fires. A considerable volume of trees suffer windthrow annually and forests composed of tall trees are more vulnerable to these influences. Secondary damage, such as insect predation, is also common after large-scale windthrow and other meteorological disturbances. **Severe weather events may also cause indirect damage to standing trees, affecting their general vigour and making them susceptible to secondary damage,** such as fungal disease and insect predation. It is estimated that 38 million hectares were disturbed annually from 2003 to 2012 by secondary effects of severe weather events (Van Lierop *et al.*, 2015).

38 million hectares were affected annually from 2003 to 2012 by secondary effects of severe weather events

→ Climatological events such as droughts or extended spells of cold weather are yet another type of disturbance to forest systems across the world. **Droughts develop slowly and can last over a year. They can occur anywhere, but their effects are particularly critical for rain-fed agriculture in semi-arid regions.** According to projections, drought episodes are expected to increase in frequency and intensity as a result of climate change. Forests affected during 88 recorded drought episodes experienced increased tree mortality (Allen *et al.*, 2010). Floods, on the other hand, tend to cause only moderate ecological damage, since they often occur in forests particularly adapted to them. Flash floods however have a far higher impact on forests.

→ Geophysical events such as tsunamis and earthquakes can cause enormous damage when they occur, but tend to be isolated events. Earthquakes can cause elevated tree mortality and may often trigger further devastation through landslides. Tsunamis, on the other hand, suggest an interesting interplay between forests and disasters: while they pose a particular threat to forests, tsunamis' devastating effects can also be significantly reduced by the presence of mangroves.

→ Anthropogenic events such as fires, oil spills, air pollution and radioactive contamination are another serious type of threat to forest systems worldwide. Fires can quickly devastate large forest areas. They are overwhelmingly human-induced, both directly (ignition) and indirectly, i.e. through circumstances that lead to fires, including: fuel accumulation; land-management practices; and changes to land-use, such as urban development or encroachment into fire-prone landscapes.

Geophysical events		
Papua New Guinea	→	two 1935 earthquakes deforested 130 km²; the 1970 earthquake deforested 60 km².
Panama	→	shallow earthquakes in 1976 removed tree cover from 54 km².
Ecuador and Colombia	→	230 km2 of forest lost to landslides following the 1987 and 1994 earthquakes.
Colombia	→	Nevado del Ruiz's 1985 volcanic eruption killed more than 20 000 people and destroyed a forest area of unknown size.
Chile	→	the 1960 earthquake impacted and severely damaged 250 km² of temperate forest.
New Zealand	→	landslides cause significant damage to forest areas.

→ **Indonesia** – Wildfires recur with enormous impacts on human health and well-being, trade and transport. Fires in Indonesia have constrained GDP growth and cost an estimated IDR 221 trillion – more than twice the reconstruction cost of the 2004 Aceh tsunami (World Bank, 2016c). The Sumatra and Kalimantan fires of 1982–1983, 1994, 1997–1998, and 2015–2016 were internationally reported.

→ **Spain** – More than 1 350 residents were evacuated from five villages in Aragon in May and December 2015 due to fires, that burned approximately 8 000 hectares of forest. The unusually high temperatures did not explain the fires' origins, according to Spain's forestry association, which said "cattle farming interests" may have been responsible for burning areas of "very high ecological value."

→ **USA** – Over 10 000 residents were evacuated and over 20 000 hectares burned in the September 2015 Valley Fire, one of the most destructive wildfires in California's history in terms of total structures burned. Overall that year, California saw four fatalities and lost 364 000 hectares to fire, at a loss of USD 1 400 million.

→ **Canada** – Driven by unseasonably high temperatures, strong winds and dryness, Alberta's May 2016 wildfire – the costliest disaster in Canadian history – covered 590 000 hectares and destroyed approximately 2 400 homes and buildings.

→ **United Republic of Tanzania** – Between 10 and 14 percent of the land area is burned each year: 11 million hectares in a country of 88 million hectares. Protected areas, game reserves, game-controlled areas and forest reserves are highly prone to fire. This information comes from a report on burned area dynamics commissioned by the GIZ TriCo Project, and is based on 11 years of MODIS satellite data.

All these types of disasters differ in the way they cause damage to forests, the type of damage they cause and the consequences of the inflicted damage – both in terms of immediate coping mechanisms (e.g. timber salvaging) and longer-term environmental repercussions and recovery planning.

Long-term impacts and consequences of disasters on forest systems

Severe weather events can negatively affect forest environments in a variety of ways and potentially for extended periods of time. Core forest functions can be severely affected, such as providing non-wood forest products, storing carbon, ensuring water and biodiversity protection. Moreover, **damaged trees become highly susceptible to secondary agents such as fire, diseases and insect attacks that can spread into neighbouring undamaged forests**. A sequence of undesirable consequences may occur suddenly, such as heavy bamboo and/or vine infestation, and mass propagation of insects triggered by a large number of trees left lying on the ground.

→ Carbon storing: **All the different types of disasters – from hurricanes to human-induced forest fires – can significantly affect the ability of forests to store carbon.** Forest fires can result in large quantities of carbon being emitted into the atmosphere. However, forests can be regenerated and once again serve as carbon sinks. From a carbon storage perspective, it matters how timber from damaged forests is salvaged and used. If used for house construction, carbon will be sequestered for a long time, but not if used as firewood or for temporary protection (Lindenmayer *et al.*, 2008; Thorn *et al.*, 2016).

→ Biodiversity protection: Major events in fast-growing plantations, although costly, are not detrimental to biodiversity. On the contrary, events in scattered areas of natural forests can be detrimental to biodiversity as these areas may be too far away for efficient natural inseeding. If areas damaged are under protection, the reason for protecting them may be lost. Other disturbances that harm biodiversity and conservation efforts are forest-specific, i.e. they depend on where and in what type of forest the disturbance takes place.

Overall, global reporting on the types of disasters affecting forest systems and the nature and scale of their specific impacts remains extremely limited

Overall, global reporting on the types of disasters affecting forest systems and the nature and scale of their specific impacts remains extremely limited. In developing countries in particular, quantitative information on forest areas affected and the monetary value of damage caused, is either weak and sporadic or missing altogether. It is therefore crucial to work towards collecting reliable data on disaster occurrence in forestry, which will allow for a streamlined application of FAO's damage and loss assessment methodology in order to establish the significance, magnitude and severity of impacts caused by natural disasters in the forestry sector. Below is an outline of the first steps towards a sound analysis.

Assessment of disaster-related damage and loss in forestry

In the context of forestry, the impact of damage and loss differs significantly from its equivalent in the other agricultural subsectors and beyond. Practical aspects – such as the salvaging of damaged trees and their use for timber – imply that disaster impacts on forestry can also involve aspects of increased production and revenue growth. Therefore, it is crucial to develop a forestry-specific application of FAO's damage and loss assessment methodology that takes into account sector specificities. The challenges faced stem from the currently limited information and data availability, especially in developing countries.

Knowledge gaps hampering the assessment of damage and loss in forest systems

While scattered estimates of damage and loss in forestry are available, they are usually not standardized across disaster types or geographic areas. Approaches and methods for damage and loss data collection and assessment in forestry vary markedly. While in some countries major disasters are routinely and rigidly assessed, in others the sector remains largely undocumented. **Major storm damage in northern Europe, for example, is carefully monitored through targeted assessments and national forest inventories. Forest damage from cyclones in Asia, by contrast, is not systematically assessed.** The most rigorous assessments tend to occur in developed countries where existing systems of forest and land management provide a base level of data and information, such as for the storms Gudrun and Lothar in Sweden and Germany. In such cases, government systems and agencies are also able to provide the kind of emergency response that allows forest management organizations and industry to undertake assessments.

The most rigorous assessments tend to occur in developed countries, where existing systems of forest and land management provide a base level of data and information

PDNAs are designed to evaluate immediate needs for recovery and restoration and inform disaster response. Their focus, however, is on estimating socioeconomic impact; they do not currently take into account longer-term damage and loss sustained by/in forests. In addition, damage caused by small-scale fires, small windthrow events, and localized pest infestations remains largely unreported, despite meeting the universally established definition of a disaster. **Much of what we currently know about forest damage and its assessment is in the form of research reports on the application of remote sensing and does not constitute a solid basis for a rigorous sector-specific assessment.**

Nevertheless, there is good scope to enhance forestry-specific evaluation methods, including data collection and methodology application. Available PDNA-based assessments can be combined with the FAO methodological analysis proposed in this report to better assess the effects of disasters on the sector and to inform adequate rehabilitation and response. However, such an approach should remain context- and location-dependent , in order to account for the specificities of the forestry sector.

Examples of how disaster impacts on forests were assessed and reported

In the devastating case of **Hurricane Matthew in Haiti (2016)**, assessments were mainly qualitative, based on visual estimates from points accessible by road (Bloesch, 2016). In order to obtain a quantitative and objective estimate of the damage to tree cover and the volume of wood, medium- and very high-resolution satellite images were used. The use of remote sensing also allowed for monitoring of vegetation recovery using indices such as the Normalized Difference Vegetation Index (NDVI).

Overall, timely, organized and rational development of post-Matthew forest assessments were constrained by:

→   time delay between the hurricane and subsequent field mission;

→   scavenging of fallen timber by farmers, which continued for some time;

→   volume of fallen wood and its spatial distribution were not known;

→   hills that were very difficult to access.

In the joint assessment conducted after **Cyclone Nargis in Myanmar (2008)**, secondary data – provided by a range of ministries, UN agencies, past household surveys, satellite imaging, and other sources – formed the basis of damage and loss assessment. These data were validated by: field visits covering the whole delta, triangulation with the primary data collected through the Village Track Administrators, comparison to other countries' benchmarks, and by consultations with communities and local stakeholders. Due to limited access to centralized damage data, analyses relied on Government-supplied figures.

The systematic assessment of forests undertaken following **Storm Gudrun in Sweden (2005)** used existing forest inventory and supporting information, including growing stock volumes, age-class distribution, and species composition. The Swedish Forestry Agency collaborated with research institutions and benefited from their significant data, forest information, skills and capacities to assess damage to and loss of forests.

In the absence of existing suitable methods, **FAO has developed a special component for damage and loss assessment within the forestry sector**, which is currently integrated in its broader disaster impact assessment methodology (as presented in the Annex of this report). As far as possible, the proposed forestry-specific component takes into account the specific aspects of the sector when evaluating impact. However, this only forms the basis of a solid assessment of disaster damage and loss in forestry. It should be complemented and enriched by identifying, collecting and streamlining the data required to test and apply this methodology.

In order to achieve **sub-regional aggregations and conduct consistent analyses**, further efforts are required on key issues, such as threshold values for assessment work, and type of volume to be assessed. Threshold issues that need to be settled include volume over or under bark, whether branches are to be included in damage statistics, where to measure the diameter on snapped trees and how to include the standing parts of those trees.

Robust **methods to assess environmental effects of disasters** also need to be developed. This would require setting up an analytical framework, populating it with any existing methods or potential methods from other disciplines or disaster assessment processes, and identifying the data gathering gaps that exist. This step will provide the basis for a targeted review of methods related to data gaps, with a view to adapting those that are relevant and can support data gathering.

A systematic process must be set in place to collect the data required. It is possible to draw from existing national work to enable comparisons between events and to develop definitions of damage types, as well as post-disaster issues to resolve

One lesson from Hurricane Matthew, Cyclone Nargis and Storm Gudrun is that the **quality, consistency and clarity of existing data sets** and documentation (methodologies) is as important as their availability. Such data sets and methodologies were present for Storm Gudrun, but not for Hurricane Matthew or Cyclone Nargis. Continuous improvement of forest data collection, forest monitoring and systems for recording, storing and analysing forest data is critical, should be an ongoing process in all countries and receive greater investment in developing ones. A systematic process must be set in place to collect the data required and further fine-tune the methodology. It is possible to draw from existing national work to enable comparisons between events, and to develop definitions of damaged trees, damage types, and post-disaster issues to resolve.

Remote sensing can be useful in determining the size, location and distribution of affected areas. Helicopters and drones can be used for rapid assessment to support initial relief efforts. While areas affected by fire can be identified on satellite images, more detailed studies require aerial photos, for which drones are a promising tool. Though combinations of remote sensing and field measurements have been used (e.g. Bjorheden, 2007), no comprehensive guidance exists on how to best integrate these methods.

Towards understanding the effects of disasters on forestry: next steps

Given the specificity of the forestry sector it is important to develop a tailored approach to analysing disaster damage and loss

Given the specificity of the forestry sector in terms of how it interacts with various types of disasters, it is important to develop a tailored approach to analysing disaster damage and loss in the forestry context. While FAO proposes a methodology that meets the basic criteria, it still remains to be trialled and tested for the subsector. Meanwhile, the challenge ahead for developing countries and the international DRR community is to develop a solid pool of data on impacted forest areas, in terms of their size, volume of damaged trees, salvaged timber, impact on associated livelihoods, etc. It is crucial to build stronger capacities for data collection and information management and to enhance technical understanding of relevant assessment methodologies among key sector stakeholders. **Only when a comprehensive information system is in place can the methodology be applied to evaluate the nature, size and magnitude of the impact of disasters on different forest systems across regions.** Such an understanding is crucial to inform adequate policy decisions and allow for an effective and holistic monitoring of agreed international resilience targets under the Sendai Framework and the SDGs. Securing a place for forestry on the map of global resilience-building is a critical challenge ahead.

Furthermore, it is important to consider the enormous potential of forests and forest environments to act as buffers and protectors against natural hazards and disasters. In the context of post-disaster damage and loss assessments, forests (including damaged ones) can provide a range of immediate uses, benefits and tools for disaster prevention and DRR in the affected areas and communities. This effort requires key information: location and distribution of areas affected, area and volume, and assessment of the damaged timbers' usefulness. While no general methods or recommendations yet exist for such assessments, the section on mitigation through forest conservation takes a more detailed look at the role of forests in DRR and disaster impact mitigation.

Mitigation through forest conservation: the role of forests in reducing vulnerability to natural disasters

Forests should not only be seen as victims of natural disasters – evidence from around the world points to the important role of forest ecosystems in mitigating the impacts of natural hazards and in facilitating post-disaster reconstruction and recovery. When appropriately planned and managed, forests can not only withstand but also protect against natural disasters of varying degrees and types, including tsunamis, storm surges, cyclones, landslides, and floods. They can substantially reduce the brunt borne by communities, both in terms of limiting their physical exposure to hazards and by providing the livelihood resources necessary to recover from the crises.

Evidence from storm occurrences in the Asia-Pacific region shows that the variety of coastal forests – including beach forests, mangroves, and planted forests – offer an effective physical barrier against tidal waves, thus mitigating the effects of tsunamis and storm waves. By absorbing wave energy and stabilizing sand dunes and other elevated wave barriers, coastal forests can reduce the destructive power of tsunamis with wave heights of up to 8–10 metres. Additionally, forests can reduce secondary damage by blocking drifting objects from washing inland.

Appropriately planned forests and tree planting can also reduce flooding risks in small and medium catchments. During short-duration rainfalls, forests soils are capable of reducing runoff at small and medium scales due to their enhanced interception, infiltration and storage capacities, as compared to grass or bare lands.

While forests can provide a measure of protection to vulnerable communities, mounting pressure on forest land and resources, coupled with poor governance and rural poverty, can pose serious challenges. In order to ensure forestry's role in effective disaster impact mitigation and reconstruction, a number of policy implications should be acknowledged. Smart forest planning and sustainable resource management are necessary to ensure disaster-resilient coastal zone development. **Managing forests for disaster protection should aim to minimize the impacts of disasters by implementing a variety of sustainable measures that work with natural processes.** Therefore, forest conservationists need to better appreciate that reducing poverty, ensuring local forest rights and promoting good governance are not only essential components of any disaster reduction strategy but are also critical for long-term biodiversity conservation. Even though the linkages are often complex, the basic building blocks of disaster reduction and forest conservation strategies are the same.

Overall, forestry can make a strong contribution to disaster risk management, especially when combined with appropriate land-use planning, mitigation measures and early warning systems. This underscores the fact that integrated approaches to DRR are the most effective. Rather than focusing solely on disaster prevention, they should aim equally at disaster mitigation and adaptation.

Evapotranspiration from forests is also an effective tool to reduce soil moisture content below that achieved by other vegetation types, thus creating a more substantial buffer against flooding during rainfall. However, forests are less effective when it comes to mitigating more severe, basin-wide events and downstream flooding.

Trees and forests can serve to prevent landslides and soil erosion on hillsides and mountain slopes. Deep-rooted trees and shrubs can reinforce shallow soil layers, anchor soil to bedrock, and form buttresses that resist soil movement. Forest cover and undergrowth also reduces soil moisture levels, which reduces the risk of landslides. Furthermore, dense forests constitute effective barriers to rock, debris and soil slips from higher elevations, which can limit the distance of landslide run-out. However, forest cover only has a marginal effect on deep-seated landslides that are typically instigated by extreme events such as continuous heavy rainfall or earthquakes.

Moreover, evidence has also shown the importance of forests in post-disaster recovery, by providing timber for home and infrastructure reconstruction. Wood is an important building material in post-disaster reconstruction and given its low weight and high strength-to-weight ratio, it can be particularly useful in earthquake-prone areas. Wood from damaged forests can be salvaged and used for immediate relief in the form of heat, shelter, firewood, etc. In the a longer term, wood can be utilized to reconstruct infrastructure and energy needs for societies.

Wood energy for strengthened resilience of displaced people

The global population of displaced people has risen to more than 65 million, compared to 37.5 million a decade ago, and a large proportion now live in overflowing displacement settings where natural resources are often limited. Of the displaced population, 80 percent rely on traditional biomass fuels, particularly fuelwood and charcoal for cooking and heating. It is expected that wood will remain a primary source of energy for the foreseeable future.

This poses multiple challenges: food security is threatened, women are endangered by leaving the camps to scavenge for fuelwood, and the environment of host communities is placed at risk as forests and other woodlands are degraded and destroyed. The situation is well explained in FAO's 2017 **video** on the Gambella, Ethiopia camps for displaced South Sudanese (available at https://www.youtube.com/watch?v=ysIoC6vO4A4&feature=youtu.be).

Displacement settings tend to have a much longer life-span than expected upon construction, with most lasting more than 20 years (UNHCR, 2016). Plans for long-term sustainable fuel supplies and management of natural resources are usually not made when camps are opened.

FAO and UNHCR have developed a methodological basis for assessing woodfuel supply and demand in displacement settings. The two agencies are working together to build resilience in crises while promoting sustainable management of forests for the benefit of the local environment, communities, and displaced people. Upon the construction of displacement camps, the following steps should be considered:

→ **Make a baseline assessment** – from what areas can fuelwood be sourced, what is the potential sustainable annual harvesting volume? This analysis will often show that more wood must be grown.

→ **Develop sustainable forest-management plans**, including afforestation/reforestation with fast-growing tree species for energy sourcing, as well as other multipurpose plantations for ecosystem rehabilitation. This provides a key contribution to reducing the environmental impact, enhancing the supply of woody biomass and non-wood forest products and responding to climate change mitigation and adaptation needs.

→ **Engage both host and displaced communities in forest management.** For instance, employing camp residents on forest plantations and the surrounding areas provides productive engagement and helps host communities manage their forest resources, improving livelihoods for both. This has multiple positive implications at the social and economic levels, and may greatly decrease the risk of tensions.

→ **Promoting fuel-efficient cookstoves and energy-saving** measures brings demand down to more sustainable levels and contributes to more sustainable livelihood opportunities.

The Bidibidi refugee settlement in northern Uganda

Uganda is currently hosting over one million refugees who have fled from war, drought and famine in the neighbouring countries of South Sudan, Burundi and the Democratic Republic of the Congo (DRC). The recent influx of refugees from South Sudan has prompted one of the most severe humanitarian emergencies in Uganda and led to the establishment of the Bidibidi settlement in the Yumbe District in August 2016. **This rapidly expanded to become the largest refugee-hosting area in the world, with 272 206 refugees over a total assigned area of 798 km². Bidibidi comprises more than half the population of its host district Yumbe (484 822 people).** The need to meet growing settlement demands for woodfuel for cooking and heating has resulted in increased felling of trees around the area, putting tremendous pressure on the environment.

Furthermore, the Bidibidi settlement refugees are living in a status of acute emergency and are exposed to a wide array of vulnerabilities. With their livelihoods disrupted, refugees have no source of income and lack secure access to fuel and energy for cooking. This triggers other vulnerabilities including compromised nutrition, food security, health and safety. Meanwhile, the wider community is affected by issues of tension/conflict and potentially irreversible environmental degradation. Some of these impacts can be reduced through participatory planning, implementation and monitoring.

A rapid woodfuel assessment was jointly initiated by UNHCR and FAO in March 2017 to assess the supply and demand of woodfuel resources in the area. It improves understanding of the acute emergency and associated environmental impacts, providing a basis for planning and monitoring strategic interventions to sustainably manage forest resources and improve energy access. The methodology included three phases:

→ assessment of the woodfuel demand;
→ assessment of the woodfuel supply;
→ identification of relevant inter-linkages, gaps, opportunities and alternative scenarios.

The rate of fuelwood demand in the Bidibidi is estimated at 347 480 tonnes per year, while the total above ground biomass (AGB) stock in the settlement area is currently 734 614 tonnes with an annual AGB growth of 33 300 tonnes per year. Assuming a constant fuelwood demand, the annual AGB loss is 314 180 tonnes. In a worst-case scenario, the total AGB stock and growth from trees and shrubs within the settlement area can supply fuelwood to the Bidibidi for up to three years. This, however, would fully deplete AGB in the area, an issue that requires urgent attention.

Continuous monitoring of these resources is needed to inform decision making and formulate measures to mitigate pressure and support sustainable energy access for refugees. A combination of immediate interventions is needed to reduce high woodfuel demand by the population of the Bidibidi settlement and to support a sustainable woodfuel supply. These include:

→ introducing and promoting fuel-efficient cookstoves and energy-saving practices at the household level;
→ establishing multipurpose tree planting programmes for building resilience and creating opportunities for sustainable livelihoods;
→ developing community-based forest resources management plans;
→ promoting woodfuel alternatives (e.g. agricultural residues, liquefied petroleum gas [LPG]).

Such interventions would have the added value of contributing to climate change mitigation through carbon sequestration and reduction of CO_2 emissions. The carbon sequestrated by forest plantations and the emission reductions from fuel-efficient cook stoves and energy-saving practices could provide additional benefits through carbon credits.

PART III ⟨ Covering new ground:
food chain crises and conflict

South Sudan 2017 〉 Improving livestock health

Chapter VII
Food chain crises damage and loss — measuring the impact of transboundary animal diseases on livestock

The frequency and severity of animal diseases have been growing over the past decades. However, their impact on the livestock sector and the food chain remains poorly analysed and under-reported. The growing interconnectedness between natural hazards and some disease outbreaks poses further challenges. Looking at the particular cases of Bluetongue Virus (BTV) and Rift Valley Fever (RVF), this chapter leverages FAO's assessment methodology to explore the effects of animal diseases in terms of damage and loss on the livestock sector, highlighting issues to be considered when evaluating disease-related impact within the context of extreme weather events.

The breakdown on food chain crises: livestock sector perspective

Livestock play a central role in the natural resource-based livelihoods of a vast majority of the population, especially in developing countries. The sector accounts for 40 percent of the gross value of global agricultural production (FAO, 2009). Its importance is likely to grow further as rising incomes and urbanization increase the demand for livestock products.

The frequency and severity of animal disease outbreaks has been growing over the past years

However, a growing amount of livestock and livestock products are lost annually to various disease outbreaks. These can be either: 1) diseases of transboundary nature affecting livestock across borders with a negative impact on food security, nutrition and safe trade; or 2) diseases of a public health nature affecting both humans and animals (zoonotic diseases). **The human food chain is under continuous threat from such outbreaks. Every year, one in ten people falls ill from eating contaminated food, and 420 000 people die as a result.** The growing trends of livestock displacement and migration due to economic, environmental or political reasons, have intensified the spread of pathogens. From pests plaguing livestock, diseases passing from animals to humans, pollutants compromising waters and soils to climate patterns undergoing drastic change, the threats to both the livestock sector and the human food chain are ever growing.

Transboundary animal diseases (TADs) are highly contagious and can spread rapidly across vast territories, irrespective of national borders. TADs cause high morbidity and mortality in susceptible animal populations, constitute a constant threat to the livelihoods of livestock farmers, and the economic costs can often be felt at national and international levels.

Employing FAO's assessment methodology, a first attempt was made to explore the effects of TADs in terms of damage and loss and expand the focus to the entire value chain

Changing agro-ecological conditions, intensifying food production systems, and expanding global trade are among the factors that affect the likelihood of transboundary disease outbreaks and their reach. Some countries and geographic areas are more vulnerable to the spread of TADs than others, depending on their levels of economic development, political context, civil unrest, regulatory regime including resources for prevention, and their ecological and biological conditions.

Furthermore, intensified outbreaks of TADs can often be linked to the occurrence of natural hazards, which can catalyze disease-spreading conditions, affecting vector-breeding sites and vector-borne disease transmission.

While the two general types of threats – natural hazards and diseases – interact in a complex relationship, the effects of the latter on the former remain largely unexplored and are usually not taken into consideration during the PDNA process.

Both threats cause massive loss to the livestock sector as a whole, which can occur through the direct effects of a natural disaster as well as through disease incidence triggered by the latter.

Exploring the link between natural disasters and animal diseases: the interconnectedness of threats

Many diseases are linked to climate and weather events, both geographically and temporally. Temperature and humidity are important determinants of the survival of pathogens in the external environment (i.e. outside their animal host) as well as of the survival and activity of insect vectors involved in pathogen transmission. Small changes in vector characteristics can produce substantial changes in disease transmission. Higher temperatures and greater humidity generally increase the rate of development of parasites and pathogens that spend part of their life cycle outside the host, as well as shorten the interval between successive generations of vector populations. **Flooding that follows extreme rainfall can increase exposure to many water-borne pathogensing.** The combination of wind with particular atmospheric conditions has been shown to be a conduit for long-range dispersal of disease vectors and pathogens.

Animal-to-animal transmissions escalate when drought forces too many animals to gather at water ponds. Floods increase the risk of vector-borne diseases because many vectors breed in humid conditions

Droughts and floods are among the most common natural hazards that threaten agricultural systems, and both have a complex relationship with disease outbreaks. They are associated with an increased risk of water/soil-borne diseases (e.g. leptospirosis and anthrax) as the intensified use of scarce water sources during periods of drought increases the likelihood of contamination, while floods help spread water-borne pathogens over a wider area. Floods are also more likely to increase the risk of vector-borne diseases (e.g. RVF), as many vectors breed in humid conditions. On the other hand, drought is more likely to prompt an increase in the occurrence of diseases directly transmitted from animal to animal – such as foot and mouth disease (FMD) and peste des petits ruminants (PPR) (as in the Republic of Kenya in 2012) – due to animal undernourishment, reduced immunity and the congregation of diseased animals.

Table 1. Theoretical risk of acquiring infectious diseases of different modes of transmission by type of climate-related disaster/event

Type of disaster	Animal-to-animal	Vector-borne	Water/solid-borne
Drought	High	Medium	Medium
Flood	Medium	High	High

Source: adapted from Hales et al., 2003

In the longer term, extreme weather conditions can also influence disease ecology and epidemiology through changes in ecosystem structure (types and abundance of plants, vectors and wildlife) and livestock production systems. The predicted reduction in biodiversity will further diminish the ability of ecosystems to dilute disease transmission, thus intensifying the magnitude of outbreaks.

Given the interaction between them, it is often extremely difficult to disentangle the effects of climatic events from disease impacts. The recent anthrax epidemic in Siberia and the incursions of Bluetongue Virus (BTV) into Europe are examples of diseases, for which climate change/weather was unequivocally the underlying causal mechanism of impacts that were induced solely through animal diseases. These two examples illustrate the epidemiological aspects as well as the nature of damage and loss to the livestock sector from climate-induced animal diseases.

Box 1. Siberia's anthrax outbreak: how the interplay of natural hazards and disease occurrence affects the livestock sector

Anthrax, a bacterial disease, poses a major threat to livestock as well as humans worldwide. The bacterium survives in the environment as a dormant spore not requiring any nutrients. While the bacteria are themselves very resistant to climatic factors, outbreaks of anthrax tend to occur in association with heavy rainfall and flooding (which bring the bacteria to the surface), and with drought, which may induce grazing closer to the ground, and thereby lead to larger exposure to bacteria in the soil (Turner *et al.,* 2013).

In July 2016, an outbreak of anthrax occurred in nomadic reindeer in the Siberian Yamalo-Nenets region, in the arctic zone of the west Siberian plain. More than 2 300 reindeer died and 72 nomadic herders, were hospitalized, including 41 children. A state of emergency was imposed. **People were evacuated from the affected area and thousands**

Box 2. Damage and loss from unprecedented outbreak of Bluetongue virus in Europe

Table 2. Net costs (1 000 euros) of the Dutch 2006 and 2007 BTV epidemics by cost category and small ruminant (SR) type

2006	Production loss	Treatment	Diagnosis	Control	Market loss	Total
Cattle	395.0	6.1	1 772.5	16 160.4	10 086.3	28 420.3
%	1.4	0.0	6.2	56.9	35.5	100.0
SRs	69.7	3.7	493.5	3 356.2	25.0	3 948.1
%	1.8	0.1	12.5	85.0	0.6	100.0
2007						
Cattle	135 142.0	2 438.4	848.8	1 493.6	9 414.8	149 337.6
%	90.5	1.6	0.6	1.0	6.3	100.0
SRs	18 817.0	5 644.5	1 190.0	343.5	19.9	26 014.9
%	72.3	21.7	4.6	1.3	0.1	100.0

Bluetongue virus (BTV) can infect all ruminant species. It causes the most severe symptoms in improved breeds of sheep and some species of deer, though cattle are usually the main reservoir host. BTV is transmitted through certain species of biting midges and is generally restricted to tropical and sub-tropical areas, where these vector species are found.

of reindeer were vaccinated. Nevertheless, the reindeer population was reduced by 250 000 animals, a far greater reduction than in normal years.

At the time of the outbreak, an over 30°C heatwave had thawed the permafrost soil that covers much of Russia. The softening soil exposed human and animal burial grounds, which allowed the spread of bacteria, including anthrax spores from recorded epidemics of the 1940's. As people and animals have been buried in permafrost for centuries, there are concerns that rising temperatures and melting ice could see the reappearance of ancient deadly bacteria and viruses. In 1995 for example, researchers found pieces of the 1918 Spanish flu virus in corpses buried in mass graves in Alaska's tundra (Taubenberger *et al.*, 2007).

While BTV has circulated for decades in sub-Saharan Africa, Turkey and the Near East, its appearance in Europe is fairly recent. Between 1998 and 2005, six strains of BTV entered Europe more or less simultaneously and rapidly spread across 12 countries. These incursions led to the death of over 1.5 million sheep and caused major disruption to trade in livestock and livestock products. In 2006 another strain of a new African BTV serotype spread into northern Europe through an unknown route, causing over 25 000 outbreaks across five countries (Belgium, Denmark, France, Luxembourg, and the Netherlands) (Wilson & Mellor, 2009).

At the time of BTV's emergence in Europe, there had been pronounced increases in night-time and winter temperatures, fewer frost days, and changes in moisture conditions. BTV incidence in southern Europe has increased most markedly in areas where the temperature has increased by at least 1°C since the 1980s. BTV incidence in northern Europe increased in areas that have warmed by almost 1.5°C. The spatial correspondence between changes in BTV and changes in precipitation amounts is less clear, but BTV-affected areas tend to be drier on average. An increased BTV incidence in northern Europe, northern and central bands of mainland Italy, and the Balkans correspond to areas that have dried since the 1980s (Purse *et al.*, 2008).

To a large degree disease costs are determined by the adopted control strategy. During the 2006 epidemic, all ruminants in the 20 km surrounding infected farms had to be indoors at all times, which incurred significant indoor housing expenses, thus driving up control costs to around 60 percent of the total estimated cost of BTV incursion. The relaxation of the indoor housing obligation in 2007 significantly reduced the cost of control measures at the expense of higher production loss.

Table 3. Sources of damage, loss and impacts on food security attributable to natural hazards and associated diseases at household and livestock sector level

Level	Cause	Damage	Loss	Impact on food security, nutrition and the economy
Household/herd	Weather	→ Excess loss of livestock → Loss of livestock-related infrastructure	→ Distress sales and slaughter → Reduced income stream due to (i) smaller herd (ii) lower productivity induced by climatic event → Increased production costs (e.g. feed prices, water haulage, movement of stock)	→ Reduced quantity and quality of food intake → Reduced employment and income of households in upstream and downstream sectors
Household/herd	Disease	→ Excess loss of livestock caused by disease (but also interaction between weather stress and disease impact)	→ Culls for disease control → Destruction of livestock-related infrastructure → Reduced income stream due to (i) smaller herd (ii) lower productivity induced by disease → Increased production costs (e.g. treatment)	→ Reduced quantity and quality of food intake → Reduced employment and income of households in upstream and downstream sectors
Livestock sector	Weather	→ Damage to veterinary, marketing and processing infrastructure	→ Reduced upstream and downstream activity	→ Reduced growth of livestock and associated sectors
Livestock sector	Disease		→ Reduced upstream and downstream activity (movement restrictions) → Carcass disposal → Enforcement of movement restrictions → Surveillance → Vaccination	→ Reduced growth of livestock and associated sectors → Restricted access to export markets

Towards an integrated damage and loss assessment in food chain crises

The Dutch experience in the 2006 and 2007 BTV epidemics shows that estimating the cost of the outbreak throughout the livestock sector is a complex and challenging task

The growing number of outbreaks caused by existing and new emerging threats to the food chain have increased the need to better understand their impact on the agriculture sector, and on livestock in particular. The Dutch experience in the 2006 and 2007 BTV epidemics shows that estimating the cost of the outbreak throughout the livestock sector is a complex and challenging task. **Yet quantifying and assessing damage and loss associated with food chain crises is key when it comes to designing effective disease prevention, control and response mechanisms.** While FAO's damage and loss assessment methodology moves us one step closer to an integrated analysis of the impact of disasters – including disease – on the livestock sector, it is important that assessment is approached in a comprehensive and integrated manner, taking into account the interconnectedness of natural hazards, disasters and TADs, and focusing on the whole food chain. Moreover, it is important to enhance institutional collaboration in the global governance of transboundary threats to the food chain at all stages from production to consumption.

Using FAO's systematic methodology to assess the damage and loss incurred during food chain crises is imperative for governments to act quickly by taking necessary measures to prevent these threats, limit their geographic spread and minimize their impact. While the Annex provides a detailed overview of FAO's methodology, Table 3 provides a short list of items to consider when assessing damage and loss in the context of food chain crises, particularly at the interface between natural hazards, disasters, and transboundary diseases.

Climatic events and RVF outbreaks

Rift Valley Fever (RVF) is a mosquito-borne viral disease affecting both livestock and humans. Regarded as endemic to large parts of sub-Saharan Africa, RVF has repeatedly caused severe epidemics across African countries over the past 60 years and has recently expanded its range to the Saudi Arabian peninsula (Figure 1). It causes substantial morbidity and mortality among affected animals and tends to have the most severe effects on high-yielding, non-native varieties of livestock and young animals.

Historically, RVF outbreaks in Africa usually occur after three to four months of prolonged above-normal precipitation and temperatures

Between animals, RVF is transmitted mainly by mosquitoes, while humans are more commonly infected after exposure to infected animals and animal products. Historically, RVF outbreaks in Africa usually follow three to four months of prolonged above-normal precipitation and temperatures, stimulating expanded and more dense vegetation. At the opposite extreme, unusually low rainfall or drought conditions can also increase localized density of mosquito vectors by concentrating water into small pools (Anyamba *et al.*, 2012). Environmental factors such as elevated expected temperatures in the Pacific and Indian oceans, heavy rains, humidity, vegetation mass, and cloud cover – all of which help trigger larger and more vigorous mosquito populations – can serve to predict RVF outbreaks in sub-Saharan countries (Munyua *et al.*, 2010).

Moreover, shifts from El Niño to La Niña in the eastern Pacific Ocean and associated sea surface temperature anomalies in the western Indian Ocean can also be associated with shifts in RVF outbreaks and patterns (Anyamba et al., 2012). **In East Africa, over half of El Niño occurrences in recent history have been accompanied by corresponding RVF outbreaks, the 1997 RVF outbreak in Kenya being the most dramatic example.** Therefore, the high frequency of El Niño events combined with their growing severity is likely to pose serious challenges for future RVF outbreak management.

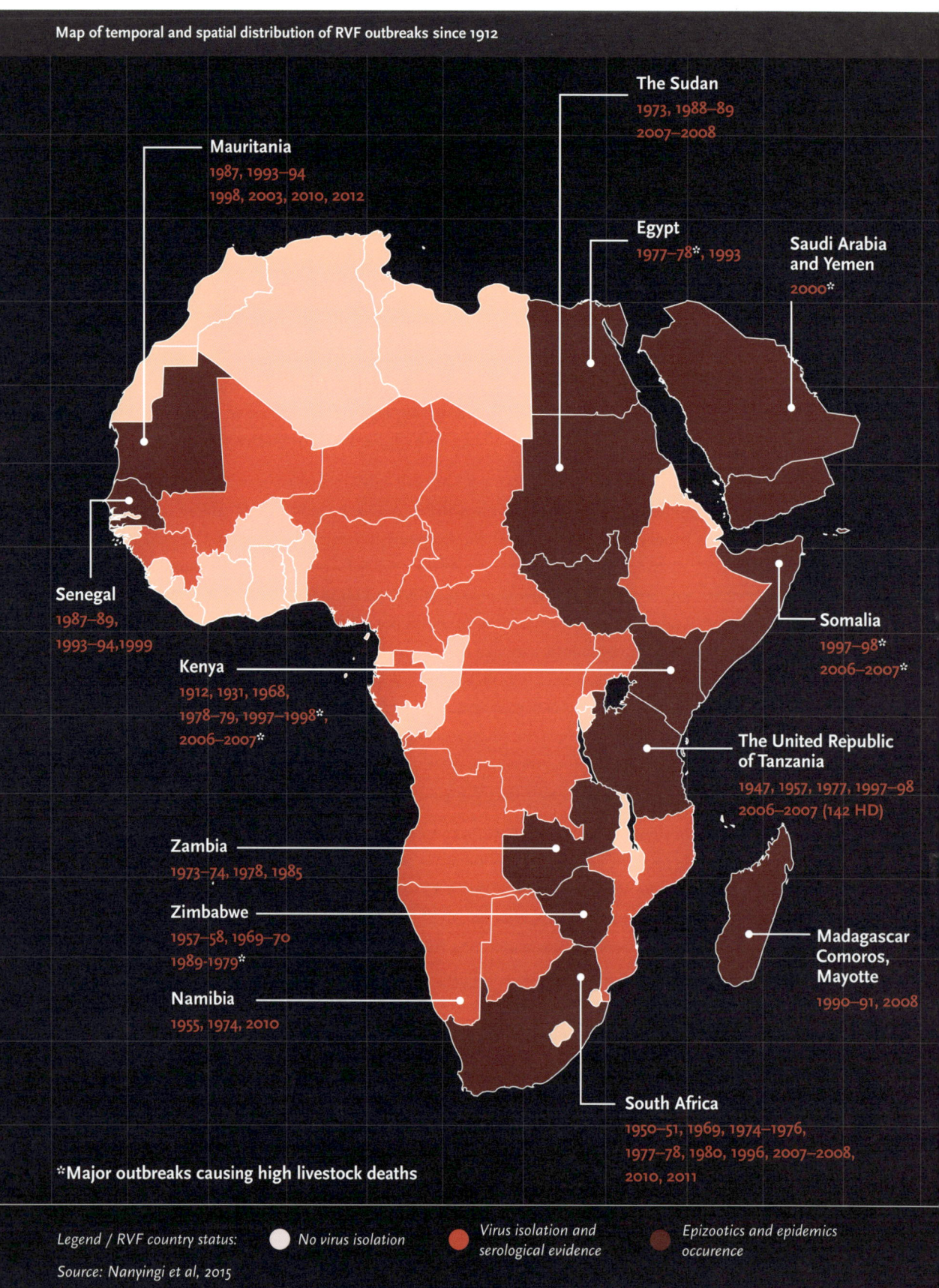

Map of temporal and spatial distribution of RVF outbreaks since 1912

The Sudan
1973, 1988–89
2007–2008

Mauritania
1987, 1993–94
1998, 2003, 2010, 2012

Egypt
1977–78*, 1993

**Saudi Arabia
and Yemen**
2000*

Senegal
1987–89,
1993–94,1999

Kenya
1912, 1931, 1968,
1978–79, 1997–1998*,
2006–2007*

Somalia
1997–98*
2006–2007*

**The United Republic
of Tanzania**
1947, 1957, 1977, 1997–98
2006–2007 (142 HD)

Zambia
1973–74, 1978, 1985

Zimbabwe
1957–58, 1969–70
1989-1979*

**Madagascar
Comoros,
Mayotte**
1990–91, 2008

Namibia
1955, 1974, 2010

South Africa
1950–51, 1969, 1974–1976,
1977–78, 1980, 1996, 2007–2008,
2010, 2011

*Major outbreaks causing high livestock deaths

Legend / RVF country status: ○ No virus isolation   ● Virus isolation and serological evidence   ● Epizootics and epidemics occurence

Source: Nanyingi et al, 2015

RVF outbreaks and their impacts

The available information on mortality rates in humans and animals and the financial impact of epidemics remain extremely limited

While a substantial number of RVF epidemics have occurred in sub-Saharan Africa since the 1950s, the available information on mortality rates in humans and animals as well as estimates of the financial impact of epidemics remain extremely limited (see Table 4). In fact, impact estimates are often limited to the value of animal loss and hardly any information is available about sources of damage and impacts on food security and nutrition.

The most comprehensive assessment of the economic impact of RVF currently available was carried out after the 2006–2007 epidemic in the Garissa and Ijara districts in northeast Kenya (Rich & Wanyoike, 2010). The RVF epidemic spread from the northeastern and coastal provinces of Kenya from December 2006 through June 2007, after anomalous heavy rains flooded large areas and caused explosive mosquito population growth. The disease spread rapidly across the predominantly pastoral areas, affecting thousands of animals. **The nomadic lifestyle of local communities further affected the spread of the outbreak, radiating the infections to surrounding areas.**

Damage and loss suffered by livestock keepers

The main negative effects on producers were caused by the loss of animals that died of RVF, which in turn had impacts on food security and future income (such as the loss of future stock caused by animal abortions). It was estimated that 420 000 animals (mainly sheep and goats) died in the two districts, representing a total value of approximately KES 610 million (around USD 9.3 million) (Table 5). Abortions in cattle and camels resulted in projected milk loss of around 2.5 million litres for a total value of close to KES 70 million (approximately USD 1.1 million). Mortality and abortions reduced flock sizes – by 22 percent for sheep – curtailing future offtake potential. For households keeping livestock, damage and loss amounted to KES 175 000 from mortality and up to KES 760 000 from loss of milk sales (Rich & Wanyoike, 2010).

The RVF epidemic severely affected food security, as 90 percent of the population was dependent on livestock for food and income

The RVF epidemic severely affected food security, as 90 percent of the population was dependent on livestock for food and income, and consumption of raw blood mixed with milk or hot soup was an important component of the diet (N'gang'a *et al.*, 2016).

Average human health costs incurred by households with a reported human case have been valued at USD 120 for every household (total estimated cost of USD 82 000 for the 2006–2007 outbreak) (Orinde *et al.*, 2012). Long-term illness and disability resulting from RVF infection impaired farmers' ability to resume their normal economic activities (Peyre *et al.*, 2015).

Damage and loss suffered by downstream value chain actors

Livestock traders were particularly impacted by the movement bans established during the RVF outbreak. Some of their animals died from RVF and most traders did not sell any animals during the outbreak. In some cases, traders incurred additional costs of maintaining animals they had purchased just before the movement bans began (Table 6). **As livestock trading typically represents about 60–80 percent of trader income, many traders were forced to rely on their savings, which made it difficult for them to resume their livestock trading activities once the outbreak was contained.**

Table 4. Human and animal cases and deaths (damage), estimated impact and applied control measures for RVF epidemics since 1950

Year	Country	Reported cases in humans	Estimated cases in humans	Confirmed deaths in humans	Reported cases in animals	Reported deaths in animals	Estimated impact (USD million)
1950 – 1951	South Africa	nd	nd	nd	600 000	100 000	nd
1977 – 1978	Egypt	nd	200 000	598	nd	nd	115
1978	Zimbabwe	nd	nd	nd	70 000	10 000	nd
1987 – 1989	Senegal	273	nd	216	1 715	nd	nd
1997 – 1998	Kenya Somalia United Republic of Tanzania	nd	89 000	478	nd	nd	250 378 nd
1998	Mauritania	300-400	nd	6	343	nd	nd
2000 – 2001	Saudi Arabia Yemen	886 1 328	20 000 nd	245 166	>10 000 22 000	1 000 6 000	10 107
2006 – 2007	Kenya Somalia United Republic of Tanzania	684 114 264	75 000 30 000 40 000	158 51 109	>4 400 nd 32 000	235 nd 4 200	41 471 nd
2007 – 2008	Sudan	747	75 000	230	nd	nd	nd
2008	Madagascar	476	10 000	19	23	18	nd
2008 – 2009	Madagascar	236	nd	7			nd
2010	South Africa	242	nd	26	>15 000	9 000	nd
2010	Mauritania	63	nd	13	nd	nd	nd
2012	Mauritania	41	nd	13	nd	343	nd

Legend: nd = no data   Sources: Dar et al., 2013; Nanyingi et al., 2015; Peyre et al., 2015

Table 5. Damage and production losses, Garissa and Ijara districts of Kenya, 2006–2007 RVF outbreak.

Damage / loss	Cattle	Goats	Sheep	Camels
Deaths (number)	36 094	135 287	223 547	26 136
Value of dead animals (KES 000)	180 472	135 287	111 774	182 951
Abortions (number)	10 510	54 773	97 302	5 370
Projected milk loss (000 litres)	656			1 960
Value of milk loss (KES 000)	9 852			58 805
Herd size reduction(%)	2	8	22	5

Source: elaborated from Rich & Wanyoike, 2010

Table 6. Financial loss of actors in the livestock production and marketing chain during the 2006–2007 RVF epidemic in Kenya

Value chain actor	Source of loss	Value of loss (KES)
Traders	→ Animals in stock that died	up to 180 000
	→ Maintenance cost of animals in stock due to sales ban	up to 21 000
	→ Loss due to decrease in animal prices	up to 24 000
Butchers	Reduced throughput or outright closure	up to 125 000
Slaughterhouses	Reduced throughput or outright closure	132 000 to 1 440 000

Source: elaborated from Rich & Wanyoike, 2010

Table 7. Changes in domestic supply from a simulated RVF shock to selected livestock subsectors

Sector	Impact (million KES)	% Change	% of overall loss
Crops	- 157.2	- 0.12	7.5
Livestock	- 178.0	- 0.34	8.5
Meat	- 107.3	- 0.16	5.2
Food	- 288.3	- 0.14	13.8
Other sectors	-1 353.8	- 0.08	64.9
Total	- 2 084.6	- 0.09	100.0

Source: elaborated from Rich & Wanyoike, 2010

Virtually all butchers and their employees were idle during the RVF outbreak, except for a few that switched to substitute products to keep their business operational. **Sales in end-markets fell by over 95 percent, from an average of 70 to 140 kg per day to just 2 to 5 kg per day. An average butcher lost between KES 76 000 –KES 125 000 (USD 1 169–USD 1 923) during the RVF outbreak in Thika (Table 6).** As with traders, butchers tried to cope with the outbreak by drawing from their accumulated savings. Similar to traders, many butchers exhausted their operating capital, which made it difficult for them to resume operations once the outbreak had been contained.

Similar to traders, many butchers exhausted their operating capital, which made it difficult for them to resume operations once the outbreak had been contained

The impact of the RVF epidemic on slaughterhouses (Table 6) varied depending on whether the slaughterhouse was inside or outside of a quarantine area. Slaughterhouses in Northeast Province and Mwingi (inside the quarantine areas) remained closed for up to three months until the slaughter ban was lifted. In both cases, the closure of slaughterhouses had impacts on a significant number of people who indirectly depend on them for their livelihoods. In Garissa, some 100 households involved in cart transport of meat and scrap sales were negatively impacted by the closure of the slaughterhouse. Approximately 60 to 80 percent of workers in Nairobi-based slaughterhouses are casual workers, many of whom were also idle during the outbreak.

Many day labourers are paid based on the volumes of meat handled, so even those that did work during the outbreak earned significantly less than normal. In the Dandora and Dagoretti slaughterhouses, incomes of day labourers fell from an average of KES 290 to KES 330 (USD 4.50–USD 5.08) per day to just KES 50 to KES 100 (USD 0.75–USD 1.53) per day.

Quantifying and assessing damage and loss associated with food chain crises is key when it comes to designing effective disease prevention, control and response mechanisms

The total loss in output to the Kenyan economy (Table 7) was estimated at KES 2.1 billion (USD 32.1 million), which amounted to less than 0.1 percent of the total national output. The combination of damage and loss from the RVF epidemic to the Kenyan economy was thus estimated at KES 2.7 billion (USD 41.4 million), of which damage accounted for 29 percent while loss propagated through various sectors of the economy accounted for the remaining 71 percent. Output loss to the livestock sector, which suffered the largest relative loss in output, was estimated at KES 178 million (USD 2.7 million) representing 8.5 percent of the total loss in output. The crop sector was estimated to have suffered a loss of KES 157 million (USD 2.4 million), most likely through a decrease in feed demand. Loss in meat and other food sectors were estimated to have been almost as high as those in the crop and livestock sectors. Though other sectors were relatively less affected, they suffered in aggregate around 65 percent of total output loss attributed to the RVF epidemic.

Syria 2016 ⟨ Southern rural Manbij

© Reuters/Rodi Said

Chapter VIII
Applying FAO's post-disaster damage and loss assessment methodology in a conflict situation – the Syrian Arab Republic.

This chapter shifts the focus to yet another type of threat – that of conflict. In the Syrian Arab Republic, FAO has just conducted the first comprehensive nationwide assessment of the impact of conflict to agriculture, adapting the new methodology in an innovative blending of primary and secondary data. The main findings provide an overview not only of the damage and loss sustained by the country's agriculture sector, but also of the support required to kick-start an effective recovery.

Natural disasters and conflicts affect millions of lives each year. In 2015 an estimated 89.4 million people were affected by natural disasters, while the number of displaced persons – an indicator of a conflict's impact – increased for the fifth time in a row to an estimated 65.3 million (Development Initiatives, 2016a). FAO estimates that 19 countries are currently experiencing protracted crises, conflict and violence (FAO, 2017c). Besides the high death tolls, both natural disasters and conflicts have a long list of potentially devastating impacts – they can bring significant economic loss and cause damage at all scales. Natural disasters and periods of protracted crises often overlap, aggravating their impacts. National economies are disrupted and public infrastructure destroyed. Affected people lose access to basic services and assets. As a result, they often face challenges in meeting their daily needs and maintaining their livelihoods.

Such has been the situation in the Syrian Arab Republic since March 2011, nearly seven years at the time of this report. This has had a devastating impact on agriculture and rural livelihoods. Yet, the sector remains a key part of the economy, contributing an estimated 26 percent to Syria's GDP, and constituting a critical safety net for the 6.7 million Syrians – including those internally displaced – who still remain in rural areas.

In 2005, 26 percent of the population was employed in agriculture (Frenken, 2009), with 38 percent of the Syrian Arab Republic's poor involved in farming, and 17 percent of all farmers living in extreme poverty compared to 11 percent of the overall population (El Laithy & Abu-Ismail, 2005). It is no surprise that agriculture and the livelihoods that depend on it have suffered massive loss throughout the conflict. Food production is currently at a record low and around half of the country's remaining population cannot meet their daily food needs. But while effective humanitarian aid and recovery planning both rely on a strong evidence base to provide a holistic and thorough picture of the damage and economic loss the war has inflicted on agriculture and its subsectors, this has been lacking. Until now.

FAO has just completed the first comprehensive nationwide assessment of the war's impact on agriculture, adapting its new assessment methodology to the specific context of conflict and protracted crises. While current PDNA guidelines go a long way towards improving the quality of assessments, including in the agricultural sector, they are best suited to assessing damage and loss from fast-onset natural disasters and are not easily applied in the context of conflicts and crises. To address this gap, and using the Syrian Arab Republic as a case study, FAO developed and tested a targeted and context-specific assessment methodology to steer a wide-ranging assessment of the Syrian agricultural sector in the context of conflict. That methodology is outlined at the end of this chapter.

Challenges to assessing damage and loss in a protracted crisis

Time is a key difference between natural disasters and conflicts. Natural disasters are usually events whose direct impacts on agriculture are generally – although not always – limited to the prevailing cropping season (drought, for example, affects multiple seasons). In contrast, conflicts often last several years and affect multiple cropping seasons. This presents the challenge of setting a reasonable baseline against which to assess damage and loss due to the conflict, and requires an estimation of the damage and loss caused during every year of conflict.

Alongside methodological challenges, conflicts and protracted crises also pose practical ones. Areas affected by conflicts are often insecure and inaccessible, even more so than in natural disasters. Conflicts also have a stronger political component, which may mean that access to some areas is restricted or even denied by authorities for political reasons. Both cases can hinder data collection. **So, while the temporal character of a protracted crisis requires collection of a comparably large amount of data to conduct a comprehensive assessment, obtaining that data is more difficult than in natural disasters.**

To achieve comprehensive coverage of the affected population and ensure that sufficient data are collected in hard-to-reach areas, an innovative solution has been found by combining various data sources and triangulating them at different levels.

Impact of the crisis on agricultural production

The Syrian Arab Republic's rural households have been hit particularly hard by the conflict

The Syrian Arab Republic's rural households have been hit particularly hard by the conflict. Vast areas of agricultural land have been destroyed and farmers are facing shortages of agricultural inputs and limited access to markets. Irrigation structures have been damaged, along with processing and storage facilities, farming equipment and buildings. **The overall financial cost of damage and loss in the agriculture sector over the 2011–2016 period is estimated by FAO to be at least USD 16 billion, equivalent to just under one-third of Syrian 2016 GDP. In terms of subsectors, annual crops registered the largest share of lost production (economic loss), followed by livestock. Conversely, the livestock subsector accounted for the highest proportion of damage (as manifested in the replacement cost of dead livestock) followed by perennial crops (as measured by the replacement cost of damaged and destroyed perennial trees).**

Annual crops

In 2007, the seven strategic annual crops in the Syrian Arab Republic (wheat, barley, cotton, sugar beet, tobacco, lentils and chickpeas) covered 75 percent of the country's cultivated land, consumed 89 percent of irrigation water and accounted for 60 percent of total crop production. The largest share of the rural population depended on cotton, rain-fed and irrigated wheat, and rainfed lentils (IFAD, 2007). As cereal production was almost fully mechanized, the largest providers of agricultural casual labour opportunities were industrial crops, such as cotton and vegetables (e.g. potatoes and tomatoes).

The challenge within the Syrian conflict context was to estimate a credible baseline for assessing annual loss for the 2011–2016 period

While the original PDNA methodology was designed to assess loss resulting mostly in the ongoing cropping season by comparing expected post-disaster production output with the latest pre-disaster levels, the challenge within the Syrian conflict context was to estimate a credible baseline for assessing annual loss for the 2011–2016 period. In order to overcome this challenge, the pre-conflict, five-year average (2006–2010) was used as the baseline production figure, assuming that this would smooth out some of the natural fluctuations in production levels that can occur over time. A period of five years was deemed a reasonable representation of crop production that could have been expected between 2011 and 2016 had there been no conflict.

This approach yielded a total loss in annual crop production of USD 4.7 billion over the course of the conflict. For the households interviewed, the area cultivated decreased by 30 percent on average, and by 50 percent for irrigated land; 10 percent had stopped crop production entirely due to insecurity and the high price of inputs. The main constraint for those households still in farming was poor access to production inputs, especially fertilizers, followed by issues related to irrigation, lack of fuel pump access, and drought.

While large annual crop loss was recorded in almost all of the Syrian Arab Republic's governorates, two stand out for a different reason: farmers in As-Sweida and Tartous experienced overall gains, illustrating that the conflict did not affect all parts of the country in a uniform way.

Permanent crops

Permanent crops accounted for 5.7 percent of land use in the Syrian Arab Republic before the crisis (ACAPS, 2013). Olive trees were by far the most cultivated, with over 90 percent of the households involved in perennial crop production reporting this either as their first, second or third main crop in terms of area. Loss in olive production and sales therefore affected almost all of the 60 percent of households still involved in perennial crop production.

Figure 1. Share of total value of damage by governorate

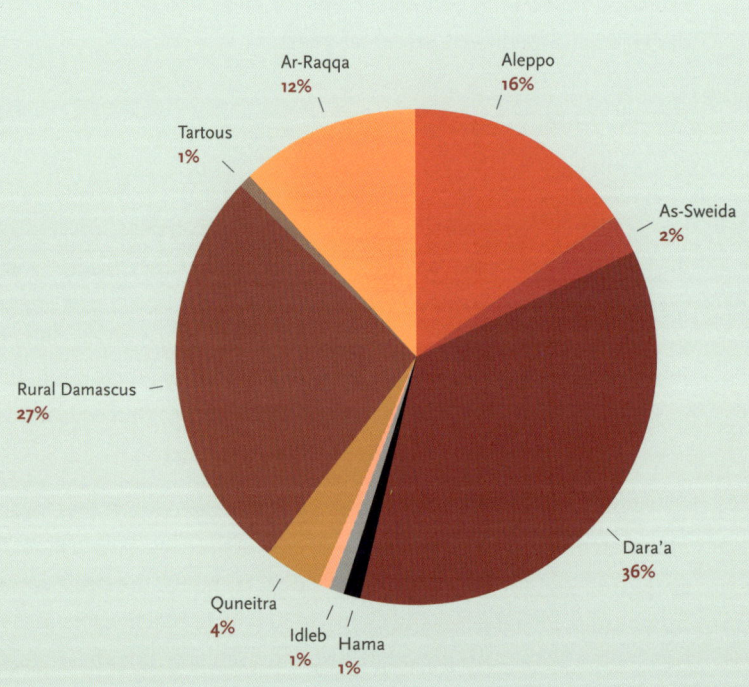

Using the same baseline period (2006–2011) and taking into account future loss until replanted perennial crops reach productive maturity (five to seven years), **loss to the perennial crop subsector was estimated at about USD 1.5 billion**. About 60 percent of households reported that lack of fertilizers was a critical production constraint for perennial crops. Lack of fuel, outbreaks of pests and diseases and lack of water resources were also listed as important constraints.

Furthermore, significant damage to tree plantations was reported in Dara'a, Rural Damascus, Aleppo and Ar-Raqqa due to destruction, but the scale is likely underestimated. **In total, the value of damage in terms of destroyed crops is estimated at around USD 903 million, equal to 13 percent of total recorded damage to the agriculture sector.**

Livestock

Livestock production played a vital role in the Syrian economy before the crisis, contributing about 35–40 percent of total agricultural production and absorbing 20 percent of rural employment. It generated approximately USD 450 million per year from exports of livestock products such as mutton, dairy, wool, leather, etc. Sheep comprised the majority of the livestock population before the crisis, while cattle and goat populations were smaller, and commercial poultry was an important source of employment. In addition, the sector contributed substantially to food security and nutrition, especially that of poor households, rural women and children in arid and semi-arid areas.

The livestock sector suffered high damage and loss amounting to USD 5.5 billion The assessment found that the livestock sector suffered high damage and loss amounting to USD 5.5 billion. In terms of destroyed assets, sector damage was estimated at a total of USD 2.7 billion, while sector loss from reduced production was around USD 2.8 billion. The greatest impact was felt in the sheep-rearing and cattle-rearing branches of the sector, where national-level damage and loss were USD 2.7 billion and USD 1.5 billion respectively. **The proportion of the shrinking rural population involved in livestock-rearing has decreased over the course of the 2011–2016 period, and the actual number of animals per household has fallen dramatically, especially for cattle.** The loss of animals, either by death due to poor living conditions, being killed or stolen was particularly high in Al-Hassakeh, Deir-ez-Zor, Lattakia, Quneitra and Rural Damascus.

The methodological approach behind these figures takes into account value estimates of crisis-related livestock deaths for the period 2010–2016, value estimates of lost animal production from dead animals, value estimates of reduced production from surviving animals (at 25 percent from original expected levels) and future loss (the value of unachieved production from new/replacement livestock).

Fisheries and aquaculture

Due to a scarcity of resources and the low natural productivity of inland fisheries, the fisheries and aquaculture sector plays only a minor role in the Syrian economy. The water area in the Syrian Arab Republic, including marshes, consists of only 1 610 sq km, representing a mere 0.9 percent of the total area of the country.

Assessing the long-lasting conflict on the fisheries and aquaculture sector relied mostly on secondary data. Assessments based on expected fisheries capture, aquaculture production and 2016 farm-gate prices indicate that the total value of damage and loss in the sector is about USD 130 million: USD 80 million in loss and about USD 50 million in damage. By far the largest reported loss was in Idleb (about USD 58 million), followed by Hama (USD 15.4 million) and Ar-Raqqa (USD 4.8 million).

Irrigation and infrastructure

Agricultural assets and infrastructure encompass a large array of elements contributing to the different agricultural subsectors. Infrastructure includes any agriculture-dedicated buildings and structures, such as cooperatives and government buildings, markets and agro-bank offices, commercial farms, veterinary clinics, animal sheds, greenhouses, storage facilities, production or transformation facilities. Assets include any tools or machinery used in agro-production, ranging from tractors, trucks and other agricultural machinery to smaller tools and equipment for post-harvest processing. Irrigation infrastructure includes canals, dams, headworks, pumps, etc.

Overall, 60 percent of households reported significant damage to infrastructure. This figure rises as high as 70–90 percent in some governorates concentrated in the most irrigated areas, i.e. Al-Hassakeh, Aleppo and Ar-Raqqa

In assessing the crisis, impact on agricultural infrastructure and assets, the methodology considered secondary data from the Economic and Social Commission for Western Asia (ESCWA, 2016), estimating the value of damage to be over USD 3.2 billion, accounting for almost half of the total damage to the agriculture sector. Overall, 60 percent of households reported significant damage to infrastructure, a figure that rises as high as 70–90 percent in some governorates concentrated in the most irrigated areas (i.e. Al-Hassakeh, Aleppo and Ar-Raqqa).

The Syrian agriculture sector relies heavily on irrigation, in particular in the northern governorates of Al-Hassakeh, Aleppo and Ar-Raqqa, as well as Deir-ez-Zor along the Euphrates. Before the crisis, some 65 percent of total cereal production was reliant on irrigation. **After six decades of increasing and inefficient use of water for irrigation, the consumption became unsustainable, placing the country under the water scarcity line.** The decrease in water resources and increased occurrence of droughts are now major concerns for the agriculture sector.

The cost of recovery

When asked what they require to enhance or resume their agricultural production, Syrian smallholders are unanimous – for annual crops, perennial crops and livestock the uniform assumption is that agricultural production can be kick-started effectively, even under current conditions. In order to do this, the emphasis should be on providing inputs (in particular fertilizer and seeds in the case of crops and feed and medicines for livestock) and on credit, marketing and processing support as well as asset repair.

The estimated costs of meeting the agricultural recovery needs expressed in household interviews and community focus groups will vary according to the scenario for conflict's foreseen evolution for the next few years. The three most likely scenarios posited by the United Nations Economic and Social Commission for Western Asia can be used to develop an indication of the possible financial implications (ESCWA, 2017).

Cost of agricultural recovery under a "partial return to peace" scenario is estimated at USD 14.9 billion

Under a "no change" scenario of the conflict continuing at its current pace, the assessment estimates that the costs over a three-year period would be in the order of USD 11 billion at 2016 prices. Under a "partial return to peace" scenario, this amount increases to USD 14.9 billion, due to an assumed partial return of rural migrants from urban areas and abroad. Under a "transition to peace" scenario, the costs amount to USD 17.1 billion.

Livestock sector recovery accounts for between 43–47 percent of total recovery costs. Annual crops account for between 29–33 percent; perennial crops between 24–26 percent

Livestock sector recovery accounts for between 43–47 percent of total recovery costs depending on the scenario, while annual crops account for between 29–33 percent, and perennial crops between 24–26 percent. The resulting estimated recovery costs per scenario are shown in Table 1.

Assuming that the next two to three years are not blighted by serious drought and/or a dramatic deepening of the crisis, ramping up investment in crop and livestock production from 2017 onwards could dramatically reduce the need for humanitarian aid, which is currently costing the international community around USD 5 billion per year. In addition, these investments could have a significant impact on internal and external migration.

Table 1. Indicative agricultural recovery costs based on farmers' preferences and needs and scenario (USD billion)

Scenario	2017–2018	2018–2019	2018–2020	Total
No change from current situation	3.56	3.56	3.56	10.68
Partial transition to peace	3.56	5.68	5.68	14.93
Transition to peace	3.56	5.68	5.68	17.05

Paving the way for agricultural reconstruction in the Syrian Arab Republic

Given the increasing number of people in need of humanitarian assistance, especially due to violent conflict and protracted crises, it is essential for the humanitarian community to be able to carry out sound agricultural sector damage and loss assessments. These can quantify the severity and magnitude of damage caused by conflict and propose costed needs-based recovery packages.

FAO has piloted an adapted PDNA-type approach that employs the formulae presented in the Annex (primary data) but also incorporates household interviews (secondary data) to provide an innovative and comprehensive assessment of agricultural damage and loss in a protracted crisis situation. It covers every year of conflict and gives a voice to the entire affected population at the governorate level, thereby providing a rich picture not only of the figures themselves but of the livelihoods behind those figures. This makes it possible to: gauge the impact of the crisis on the agricultural sector and its subsectors; identify priorities for response based on the needs and preferences expressed by the affected people, ensuring the relevance and sustainability of the associated interventions; and estimate the financial cost of meeting these priorities for resilient agricultural recovery at a national scale under different scenarios.

The methodology as applied in the Syrian Arab Republic provides an initial model to conduct more holistic needs assessments in the agricultural sector

The methodology as applied in the Syrian Arab Republic is an important step forward. It provides an initial model to conduct more holistic needs assessments in the agricultural sector for better response planning and sustainable rebuilding of more resilient livelihoods during protracted crises. After all, the only estimation of recovery costs that allows governments and FAO to "build back better" is one that treats the affected population as a valued partner, and is based on their feedback.

How damage and loss from the prolonged Syrian crisis were assessed

This first-ever damage and loss assessment of the protracted Syrian crisis on its agriculture sector was loosely rooted in the foundations of standard PDNA methodology. All basic PDNA components were considered – context analysis, setting pre-production baselines, assessing disaster effects and impacts on the sector and identifying a strategy for meeting recovery needs. However, transferring this approach to the Syrian context implied some challenges, including: 1) setting a reasonable pre-conflict baseline against which to assess conflict-related damage and loss; and 2) estimating the annual amount of damage and loss caused during the conflict. These challenges mainly arose from certain characteristics of the crisis that made the context in which the assessment was carried out different from that following a rapid-onset natural disaster, most notably the temporal dimension of protracted crises and the accessibility of the affected area.

While the original PDNA process uses pre-disaster production levels as a baseline against which to estimate economic loss, this is not applicable to the context of protracted crises. Given the time span of a conflict and the natural fluctuations in both agricultural productivity and commodity prices, the specific amount and value of production right before the onset of a crisis may not necessarily be representative of its expected amount and value in all the following years. In addition, other economic factors, such as conflict-related monetary inflation, may lead to inaccurate estimates of production loss if the value of production during or after a conflict is directly compared to its pre-conflict equivalent without adjustment. Given these issues, methodological adaptations must be made to the original PDNA process when it is transferred from a natural disaster context to a protracted crisis situation.

The assessment was carried out in two parts. Phase 1 was undertaken in August and September 2016 in 12 governorates, namely (western) Aleppo, Al-Hassakeh, As-Sweida, Dara'a, Deir-Ei-Zour, Hama, Homs, Idleb, Lattakia, Quneitra, Rural Damascus and Tartous. Phase 2 was undertaken in December 2016 and January 2017 in eastern Aleppo (controlled by Al-Nusra Front) and Ar-Raqqa (controlled by ISIS). During the field work, more than 3 500 household interviews and 383 focus group discussions (FDG) were conducted in over 380 communities. This primary data was triangulated with various secondary data sources, including estimates from the Syrian Agriculture Database (SADB) and the Central Bureau of Statistics (CBS), as well information from other assessments and documents such as the Crop and Food Security Assessment Missions (CFSAM), the Humanitarian Needs Overviews (HNOs) and the Humanitarian Response Plans (HRPs). This approach allowed for nationwide coverage and provided a picture that is both broad in terms of its geographical coverage and deep in terms of its understanding of the reasons behind the observed impacts. Additionally, the purposeful blending of primary and secondary data successfully linked the micro-level impacts of the conflict, as reflected by the affected people themselves through household interviews and focus group discussions, to macro-level impacts as deduced from secondary data.

Somalia 2017 ⟨ Building a water reservoir

Conclusion
FAO's 2017 report – Transforming damage and loss assessment

Standardizing disaster-related damage and loss assessments enables the monitoring of international targets and goals, and facilitates enhanced cooperation and coordination at the global, national and local levels. This can significantly advance progress towards more resilient and sustainable agricultural systems. The way forward is explained here.

Disasters exact a heavy toll on the agriculture sector in developing countries as they often affect production, with cascading negative consequences for national economies. The persistently growing frequency and intensity of the various hazards, threats and crises pose a serious challenge to agricultural systems. Counteracting these impacts demands adequate policy and action. To be effective, national strategies on DRR, humanitarian response, resilience and climate change adaptation must be grounded in a comprehensive understanding of the particular impact disasters have on agriculture. This has so far been impeded by significant gaps in both knowledge and data.

In 2015, FAO launched a process to improve disaster-related damage and loss information and assessment in agriculture. Its first report provided new insights on trends in damage and loss, approximations of the cost of disasters and the wider implications for livelihoods and national economies. However, the absence of a comprehensive methodology to assess damage and loss in all of agriculture's subsectors narrowed the 2015 analysis to only large-scale disasters and their impact on crops and livestock.

FAO's 2017 report: transforming damage and loss assessment

The present report goes much further than its predecessor in filling prevailing knowledge gaps, and introduces the methodology necessary to provide a holistic view of damage and loss across the whole of agriculture. It deepens the focus on crops and livestock, taking all commodities into account and considering not only large-scale events but smaller- and medium-scale disasters across developing countries, including SIDS. It also provides the first-ever analysis of the less-documented impacts borne by the forestry, fisheries and aquaculture sectors. Its expanded focus improves understanding of how agriculture is affected not only by natural disasters but by food chain crises, conflicts and protracted crises as well. This analysis tells us that:

→ **Between 2005 and 2015, approximately USD 96 billion was lost due to declines in crop and livestock production following disasters in developing countries.** About 4 percent of potential production was lost to disasters. Drought poses a major threat to both crop and livestock production and was responsible for 30 percent of overall loss.

→ While rudimentary evidence confirms the potentially large scale of damage and loss incurred by forestry, fisheries and aquaculture, no prior assessment methodology is particularly suited to these sectors. Nor do the sectors possess the systematic information base to provide the grounds for methodological post-disaster assessments. While availability of such data is limited across the board – a matter that cannot be overlooked, since building capacity for data and information management is key to designing effective disaster prevention, control and response mechanisms – nowhere is this more evident than in forestry, fisheries and aquaculture. Measuring the impact of disasters on these subsectors will be a priority of FAO's future work on damage and loss assessment.

→ Agriculture is in the crosshairs of disasters of varying kinds.

 Natural disasters: The recorded number of natural disasters in developing countries in recent years has increased almost two-fold compared to 40 years ago, and their associated impacts on rural livelihoods and agricultural economies have been staggering. **Between 2006 and 2016, crops, livestock, fisheries, aquaculture and forestry absorbed 23 percent of all damage and loss caused by medium- to large-scale natural hazard-induced disasters.**

Food chain crises: The alarming upsurge in the frequency and severity of TAD outbreaks poses a serious threat to food and nutrition security, human health, pastoral livelihoods, and overall economies. FAO's damage and loss assessment methodology brings us one step closer to an integrated analysis of the impact of food chain crises on the livestock sector and makes it possible to take into account the interconnectedness between natural hazards, disasters and TADs and the effects this has across the whole food chain.

Conflict and protracted crises: Protracted crises are the new norm, with 40 percent more ongoing food crises considered to be protracted than in 1990. FAO's adapted approach to assessing agricultural damage and loss in the context of protracted crises, as pioneered in the Syrian Arab Republic, offers a first insight into using crisis impact assessment to better inform reconstruction and humanitarian responses in agriculture. It quantifies damage, loss and recovery costs at the subnational level and blends primary and secondary data to present the devastating impact of the crisis on agriculture – **amounting to at least USD 16 billion in the Syrian Arab Republic for the 2011–2016 period.** This more-informed understanding of the country's needs and priorities fosters better response planning and sustainable rebuilding of resilient livelihoods.

In partnership with UNISDR, and as part of the common reporting guidelines initiative, FAO's methodology has already been selected to monitor progress toward the global resilience targets of international framework agreements. Specifically, it will be used to track the achievement of the Sendai Framework indicator C-2 on assessing direct agricultural loss attributed to disasters, and SDG target 1.5, which aims to build resilience and reduce exposure and vulnerability to climate-related extreme events and other shocks and disasters. Because those instruments in turn support the Paris Agreement indicators and advance the goals of the Warsaw International Mechanism for Loss and Damage Associated with Climate Change Impacts, FAO's methodology will play a key part in informing and enriching the climate change adaptation agenda.

FAO's new methodology corresponds to universal norms, commitments, collective action and shared rules at the global level. Seeking to standardize disaster impact assessment in agriculture, it is both holistic enough to be applied in different disaster events and in different country/regional contexts, and precise enough to consider all agricultural subsectors and their specificities. In addition, **it provides a framework for identifying, analysing and evaluating the impact of disasters on the agriculture sector, and constitutes a useful tool for assembling and interpreting existing information to inform risk-related policy decision-making and planning.** As the assessments of Typhoon Haiyan and the droughts in Ethiopia demonstrate, FAO's methodology can provide the backbone for damage and loss analysis in agriculture. This is true even as the methodology currently stands. It is precisely because the 2015 report highlighted the urgent need to improve data collection in every sector and at every level, that FAO designed its new methodology to accommodate varying degrees of data availability. Nevertheless, challenges lie ahead. While the foundation is there, improved data and information structures are necessary to both inform and successfully apply this methodology according to its universal potential.

Once widely adopted, FAO's methodology will be instrumental in reinforcing planning, benchmarking and accountability at the national and subnational levels. It will help to catalyze further integration of existing work on damage and loss assessment at the national level, while identifying and addressing persisting gaps and challenges in data collection. Its framework is compatible with existing national and local mechanisms and processes for post-disaster data collection (e.g. PDNAs) with broad, multi-stakeholder participation. It also widens that perspective by including climate change-induced damage and loss into the assessment framework. Therefore, **the FAO methodology offers a basis for strengthening national institutions and their statistical capacities for effective monitoring and data collection related to damage and loss caused by disasters and crises. It also emphasizes the need to foster cooperation and partnerships in support of statistical capacity-building in developing countries.**

By expanding the horizons of disaster impact assessment in agriculture and its subsectors, this report shows that FAO's methodology constitutes the all-important first step for capturing with precision the extent of disaster impact in all agro-subsectors, the food value chain, food security and overall national economies. This will help to direct investment into the agricultural sector and development assistance in a way that is commensurate with agriculture's crucial role in eradicating hunger, achieving food security and poverty alleviation and promoting sustainable development and economic growth.

The way forward: informing agricultural DRR policy

The large share of disaster impacts absorbed by agriculture, combined with the expected further increase in the frequency and intensity of natural hazards, food chain crises and conflicts, calls for enhanced and coordinated sectoral policies, actions and investments in risk reduction and climate change adaptation. The adoption of FAO's methodology to monitor the agricultural components of the Sendai Framework indicator C-2 and SDG target 1.5.2 will enable enhanced cooperation and coordination at the global, national and local levels, significantly advancing progress towards more resilient and sustainable agricultural systems.

While this study fills many information gaps, there is still much to be done.

Improving data and building knowledge on disaster impacts on agriculture – including forestry, fisheries and aquaculture – is essential

→   While FAO's damage and loss assessment methodology contributes to improved monitoring and tracking, its usefulness depends on improving local-level data in national databases and information systems, without which assessments cannot capture the full extent of disaster impacts on agriculture. This precision is fundamental for well-tailored policies and investments in the sector, and for tracking progress toward global targets – the Sendai Framework, the SDGs, the Paris Agreement and the Warsaw International Mechanism for Loss and Damage Associated with Climate Change Impacts.

Giving a voice to "silent" disasters

→   Many disasters are localized, affecting only a limited area, and often unreported. Despite their scale, such "silent" disasters exert far-reaching consequences for the livelihoods of rural communities. Taking them into account – which, again, requires improved databases – will provide a more comprehensive picture of national disaster vulnerabilities, and more targeted national DRR policy and action.

Strengthening capacity, building partnerships

→ National and subnational reporting are the most significant levels of reporting and rely heavily on the work of National Statistical Offices, National DRM Agencies and Bureaus of Agriculture. Reinforcing their capacities, mechanisms and resources for data collection, management and analysis will enable a coordinated and coherent application of FAO's assessment methodology. This in turn will build and strengthen cross-institutional partnerships, responsibility-sharing and information flow among all relevant national institutions such as Ministries of Agriculture, Forestry and Fisheries, and their departments; National Emergency Management Agencies; and National Bureaus of Statistics. Enhanced capacity for damage and loss assessment ultimately means better informed policy, action and improved DRR, preparedness and resilience in agriculture.

Streamlining damage and loss assessment efforts for better DRR policy, improved resilience and higher investment in agriculture

→ Efforts to measure disaster impact on agriculture, including FAO's framework for damage and loss assessment, should be fully integrated into overall national resilience and DRR policy, planning and implementation.

→ Given the high impact absorbed by the sector and the staggering amount of loss incurred, more financial resources should be directed to agriculture in developing countries. National governments, the private sector and international development agencies share responsibilities in terms of DRR investment.

→ National governments and the international community should establish targets for financing DRR in agriculture to prevent and mitigate the impact of disasters.

FAO works at the nexus of disasters, agriculture and climate change adaptation. Its new damage and loss methodology – capable of assessing the agricultural impacts of climate-induced extreme events – is key to meeting future challenges. It promotes strengthening technical and institutional capacities for data collection, monitoring and reporting – capacities that must be systematically embedded into sector-specific national agriculture development plans and investments. If consistently adopted and continuously fine-tuned, this methodology supplies a crucial tool for better planning and delivery of resilience, as well as DRR and management. This is particularly important for countries facing recurrent disasters, and those where agriculture is a critical source of livelihoods, food security and nutrition, as well as a key driver of economic growth.

APPENDICES

Annex

FAO's Damage and loss computation methodology Rationale, background and context

Detailed assessments of economic loss and damages are regularly carried out by governments and multilateral organizations following large-scale disasters using different methodologies. However, when applied to agriculture, **these assessments often fail to capture the specificities of the sector and result in an imprecise or under-estimated evaluation of disaster impact**. Moreover, given the lack of a universal assessment methodology, disaster impact tends to be estimated based on variations of either PDNA or Economic Commission for Latin America and the Caribbean (ECLAC) -derived methodologies, making it impossible to compare results across countries or disasters. **It is often difficult to determine which methodology, criteria and parameters have been used and which elements of agricultural damage and loss have been considered.**

Aiming for a standardized approach to assessing disaster damage and loss in agriculture, **FAO has developed a methodology that is both holistic** enough to be applied in different disaster events and in different country/regional contexts, **and precise** enough to consider all agricultural subsectors and their specificities. In addition, a common streamlined methodology can help address the prevailing knowledge gap on disaster impact on the sector and provide a useful tool for assembling and interpreting existing information about both past and future events.

Since its development, the FAO methodology has been applied in two case studies, aiming to quantify the effects of the 2013 Typhoon Haiyan on the agricultural sector in the Philippines, and of the 2008/2011 drought on the crops and livestock sectors in Ethiopia (Chapter 3). **The consistent and thorough results obtained from testing the methodology i**n the context of two typologically different disaster events and in two different country contexts with varied degrees of data availability, **have confirmed its relevance.** Elements of this methodology are also used to assess crop damage following the 2012 earthquakes in Nepal (Chapter 4) as well as the cost of conflict for the Syrian agricultural sector (Chapter 8).

FAO's methodology to assess direct economic impact of disasters on agriculture has recently been incorporated in the framework for monitoring progress in achieving the global targets of the Sendai Framework for Disaster Risk Reduction (SFDRR). In a 2017 Resolution (A/RES/71/276), the United Nations General Assembly endorsed the Report of the Open-ended Intergovernmental Expert Working Group on Indicators and Terminology Related to Disaster Risk Reduction and their recommendations for indicators and terminology. As the custodian agency, the United Nations Office for Disaster Risk Reduction (UNISDR) was responsible for developing these methodological indicators in collaboration with relevant technical partners. Within this process, FAO has worked closely with UNISDR to fine-tune and adapt the Damage and Loss Assessment Methodology. The latter is now endorsed within the proposed SFDRR monitoring framework and will serve to measure indicator C-2 (direct agricultural loss attributed to disasters).[1]

Furthermore, as part of the common reporting guidelines initiative, this methodology will also be integrated into the indicators used to track progress towards achievement of SDG target 1.5,[2] which aims to build resilience and reduce exposure and vulnerability to climate-related extreme events and other shocks and disasters.

---

1   SFDRR indicator C-2: Direct agricultural loss attributed to disasters. It measures the monetary damage to agricultural assets and infrastructure, as well as the value of production loss attributed to disasters in the crops, livestock, fisheries, aquaculture and forestry sectors. This indicator is part of a compound indicator that measures progress towards reducing direct disaster economic loss in relation to global gross domestic product by 2030.

2   SDG target 1.5: By 2030, build the resilience of the poor and those in vulnerable situations and reduce their exposure and vulnerability to climate-related extreme events and other economic, social and environmental shocks and disasters.

Overview of methodology for damage and loss assessment

FAO's methodology uses a standardized computation method to assess the direct damage and loss that occurs in the agricultural sector as a result of disasters, which takes into consideration the specificities of each subsector, i.e. crops, livestock, forestry, aquaculture and fisheries.

This computation method aims to measure the direct effects of a broad range of disasters of different types, duration and severity. Moreover, it applies to a range of disasters – from large-scale shocks to small- and medium-scale events with a cumulative impact.

The methodology consists of five composite indicators

→ **DL (C):** [Direct] damage and loss to crops

→ **DL (L):** [Direct] damage and loss to livestock[3]

→ **DL (FO):** [Direct] damage and loss to forestry

→ **DL (AQ):** [Direct] damage and loss to aquaculture

→ **DL (FI):** [Direct] damage and loss to fisheries

In combination, these indicators aim to capture the total effect of disasters on the agriculture sector:
Impact to Agriculture= DL(C)+DL(L)+DL(FO)+ DL(AQ)+DL(FI)

Key concepts

→ **Concept 1: Damage and loss**

In order to capture the direct impact of disasters on agriculture, it is important to take into account both the damage and the loss accrued in the sector. **Damage** is defined as the replacement and/or repair cost of totally or partially destroyed physical assets and stocks in the disaster-affected area. **Loss** refers to changes in economic flows arising directly from the disaster and accrued within the agricultural cycle coinciding with the disaster (this includes declines in output in crops, livestock, fisheries, aquaculture and forestry).

→ **Concept 2: Production and assets**

Each subsector is divided into two main sub-components, namely **production and assets**. The production component measures both damage and loss from disaster on production inputs and outputs, while the assets component measures damage on facilities, machinery, tools, and key infrastructure related to agricultural production.

---

3   May also include apiculture.

Main components and formulas

The table below describes the key components of the damage and loss assessment methodology, including an indication of the items that should be considered in the assessment of each sub-sector. Furthermore, this section presents the calculation methods used for assigning a monetary value to both damage and loss. A detailed presentation of the subsector-relevant formulas is provided for each damage and loss indicator.

Production			
Damage		**Loss**	
Items	**Measurement**	**Economic flows**	**Measurement**
Stored inputs Seeds, fertilizer, feed, fodder, etc. **Stored production** Crops, livestock produce, fish, logs, etc. **Perennial trees**	1. Pre-disaster value of destroyed stored production and inputs	Value of lost production (excluding stored outputs)	1) Difference between expected and actual value of production (crops, livestock, forestry, aquaculture production and fisheries capture) in disaster year **For perennial crops and forestry:** 2) Pre-disaster value of fully destroyed standing crops and trees **For crops, livestock and aquaculture:** 3) Temporary costs incurred towards the maintaining of post-disaster agricultural and farming/fishing activities
Assets			
Items	**Measurement**	**Economic flows**	**Measurement**
Machinery, equipment and tools[4] used in crop and livestock farming, forestry, fisheries, aquaculture, apiculture	**Total destruction:** replacement cost of fully destroyed assets at pre-disaster price **Partial destruction:** repair/rehabilitation cost of partially destroyed assets at pre-disaster price		

---

4   Includes (but is not limited to): tractors, balers, harvesters and threshers, fertilizer distributors, ploughs, root or tuber harvesting machines, seeders, soil machinery, irrigation facilities, tillage implements, track-laying tractors, milking machines, dairy machines, machinery for forestry, wheeled special machines, portable chainsaws, fishing vessels, fishing gear, aquaculture feeders, pumps and aerators, aquaculture support vessels, etc.

1. DL-C – Damage and Loss in Crops

DL-C (Crop damage and loss) = Annual crop production damage + Perennial crop production damage + Annual crop production loss + Perennial crop production loss + Crop assets damage (complete and partial)

→ **1.1 Annual Crop Production Damage is composed of the:**

1) Pre-disaster value of destroyed stored inputs: $\Delta q_{x(stored)j,t} \cdot p_{x(stored)j,t-1}$

2) Pre-disaster value of destroyed stored annual crops: $\Delta q_{i(stored)j,t} \cdot p_{i(stored)j,t-1}$

$$PD(AC)_{ij} = \Delta q_{i(stored)j,t} \cdot p_{i(stored)j,t-1} + \Delta q_{x(stored)j,t} \cdot p_{x(stored)j,t-1}$$

→ **1.2 Perennial Crop Production Damage is composed of the:**

1) Pre-disaster value of destroyed stored inputs: $\Delta q_{x(stored)j,t} \cdot p_{x(stored)j,t-1}$

2) Pre-disaster value of destroyed stored perennial crops: $\Delta q_{i(stored)j,t} \cdot p_{i(stored)j,t-1}$

3) Replacement value of fully damaged trees: $\Delta ha_{ij,t} \cdot h_{ij} \cdot p_{h\,j,t-1}$

$$PD(PC)_{ij} = \Delta q_{i(stored)j,t} \cdot p_{i(stored)j,t-1} + \Delta q_{x(stored)j,t} \cdot p_{x(stored)j,t-1} + \Delta ha_{ij,t} \cdot h_{ij} \cdot p_{h\,j,t-1}$$

→ **1.3 Annual Crop Production Loss is composed of the:**

1) Difference between expected and actual value of crop production in non-fully damaged harvested areas:

2) Pre-disaster value of destroyed standing crops in fully-damaged areas: $p_{ij,t-1} \cdot y_{ij,t-1} \cdot \Delta ha_{ij,t}$

3) Short-run post-disaster maintenance costs (expenses used to temporarily sustain production activities immediately post-disaster): $P_{short-run}$ (lump-sum)

$$PL(AC)_{ij} = p_{ij,t-1} \cdot \Delta y_{ij,t} \cdot ha_{ij,t} + p_{ij,t-1} \cdot y_{ij,t-1} \cdot \Delta ha_{ij,t} + P_{short-run}$$

→ **1.4 Perennial Crop Production Loss is composed of the:**

1) Difference between expected and actual value of crop production in non-fully damaged harvested areas: $p_{ij,t-1} \cdot \Delta y_{ij,t} \cdot ha_{ij,t}$

2) Pre-disaster value of destroyed standing crops in fully-damaged areas: $p_{ij,t-1} \cdot y_{ij,t-1} \cdot \Delta ha_{ij,t}$

3) Short-run post-disaster maintenance costs (expenses used to temporarily sustain production activities immediately post-disaster): $P_{short-run}$ (lump-sum)

$$PL(PC)_{ij} = p_{ij,t-1} \cdot \Delta y_{ij,t} \cdot ha_{ij,t} + p_{ij,t-1} \cdot y_{ij,t-1} \cdot \Delta ha_{ij,t} + P_{short-run}$$

→ **1.5 Crops Assets Damage is composed of the:**

1) Repair / replacement cost of partially / fully destroyed assets at pre-disaster price: $p_{kj,t-1} \cdot \Delta q_{kj,t}$

$$AD(ALL)_{ij} = p_{kj,t-1} \cdot \Delta q_{kj,t}$$

2. DL-L – Damage and Loss in Livestock

DL-L (Livestock damage and loss) = Livestock production damage + Livestock production loss + Livestock assets damage (complete and partial)

→ **2.1 Livestock Production Damage is composed of the:**

1) Pre-disaster value of stored inputs (fodder and forage): $\Delta q_{x(stored)j,t} \cdot p_{x(stored)j,t-1}$

2) Pre-disaster value of destroyed stored animal products: $\Delta q_{i(stored)j,t} \cdot p_{i(stored)j,t-1}$

3) Pre-disaster net value of dead animals: $(\Delta q_{ij,t} \cdot w_i)\cdot(p_{ij,t-1} - \alpha \cdot p_{ij,t})$

$$PD(L)_{ij}=\Delta q_{x(stored)j,t} \cdot p_{x(stored)j,t-1}+\Delta q_{i(stored)j,t} \cdot p_{i(stored)j,t-1} + (\Delta q_{ij,t}\cdot w_i)\cdot(p_{ij,t-1}-\alpha\cdot p_{ij,t})$$

→ **2.2 Livestock Production Loss is composed of the:**

1) Difference between expected and actual value of production (of livestock products): $q_{ij,t} \cdot p_{ij,t-1}\cdot\Delta y_{ij,t}$

2) Short-run post-disaster maintenance costs expenses used to temporarily sustain production activities immediately post-disaster): $P_{short-run}$ (lump-sum)

$$PL(L)_{ij}=(q_{ij,t} \cdot p_{ij,t-1}\cdot\Delta y_{ij,t})+P_{short-run}$$

→ **2.3 Livestock Assets Damage is composed of the:**

1) Repair/replacement cost of partially/fully destroyed assets at pre-disaster price: $p_{kj,t-1}\cdot\Delta q_{kj,t}$

$$AD(ALL)_{ij}=p_{kj,t-1}\cdot\Delta q_{kj,t}$$

3. DL-FO – Damage and Loss in Forestry

DL-FO (Forestry damage and loss) = Forestry production damage + Forestry production loss + Forestry assets damage (complete and partial)

→ **3.1 Forestry production damage is composed of the:**

1) Pre-disaster value of stored inputs: $\Delta q_{x(stored)j,t} \cdot p_{x(stored)j,t-1}$

2) Pre-disaster value of destroyed stored products: $\Delta q_{i(stored)j,t} \cdot p_{i(stored)j,t-1}$

3) Replacement value of fully damaged trees: $\Delta ha_{ij,t}\cdot h_{ij}\cdot p_{h\,j,t-1}$

$$PD(FO)_{ij}=\Delta q_{x(stored)j,t} \cdot p_{x(stored)j,t-1}+\Delta q_{i(stored)j,t} \cdot p_{i(stored)j,t-1} + \Delta ha_{ij,t} \cdot h_{ij} \cdot p_{h\,j,t-1}$$

→ **3.2 Forestry Production Loss is composed of the:**

1) Difference between expected and actual value of production in non-fully damaged harvested area: $ha_{ij,t} \cdot p_{ij,t-1} \cdot \Delta y_{ij,t}$

2) Pre-disaster value of fully destroyed standing forest products: $\Delta ha_{ij,t} \cdot p_{ij,t-1} \cdot y_{ij,t-1}$

$$PL(FO)_{ij}=\Delta ha_{ij,t} \cdot p_{ij,t-1} \cdot y_{ij,t-1} + ha_{ij,t} \cdot p_{ij,t-1}\cdot\Delta y_{ij,t}$$

→ **3.3 Forestry Assets Damage is composed of the:**

1) Repair / replacement cost of partially / fully destroyed assets at pre-disaster price: $p_{kj,t-1} \cdot \Delta q_{kj,t}$

$$AD(ALL)_{ij}=p_{kj,t-1} \cdot \Delta q_{kj,t}$$

4. DL-AQ – Damage and Loss in Aquaculture

DL-AQ (Aquaculture damage and loss) = Aquaculture production damage + Aquaculture production loss + Aquaculture assets damage (complete and partial)

→ **4.1 Aquaculture Production Damage is composed of the:**

1) Pre-disaster value of stored inputs: $\Delta q_{x(stored)j,t} \cdot P_{x(stored)j,t-1}$

2) Pre-disaster value of destroyed stored aquaculture products: $\Delta q_{i(stored)j,t} \cdot P_{i(stored)j,t-1}$

3) Pre-disaster net value of dead fish (brood stock loss): $(\Delta q_{ij,t} \cdot W_i)$

$$PD(AQ)_{ij} = \Delta q_{x(stored)j,t} \cdot P_{x(stored)j,t-1} + \Delta q_{i(stored)j,t} \cdot P_{i(stored)j,t-1} + (\Delta q_{ij,t} \cdot W_i)$$

→ **4.2 Aquaculture Production Loss is composed of the:**

1) Difference between expected and actual value of aquaculture production in non-fully damaged aquaculture areas: $area_{ij,t} \cdot p_{ij,t-1} \cdot \Delta y_{ij,t-1}$

2) Pre-disaster value of aquaculture production lost in fully damaged aquaculture areas: $\Delta area_{ij,t} \cdot p_{ij,t-1} \cdot y_{ij,t-1}$

3) Short-run post-disaster maintenance costs (expenses used to temporarily sustain production activities immediately post-disaster): $P_{short-run}$ (lump-sum)

$$PL(AQ)_{ij} = \Delta area_{ij,t} \cdot p_{ij,t-1} \cdot y_{ij,t-1} + area_{ij,t} \cdot p_{ij,t-1} \cdot \Delta y_{ij,t-1} + P_{short-run}$$

→ **4.3 Aquaculture Assets Damage is composed of the:**

1) Repair / replacement cost of partially / fully destroyed assets at pre-disaster price: $p_{kj,t-1} \cdot \Delta q_{kj,t}$

$$AD(ALL)_{ij} = p_{kj,t-1} \cdot \Delta q_{kj,t}$$

5. DL-FI – Damage and Loss in Fisheries

DL-FI (Fisheries damage and doss) = Fisheries production damage + Fisheries production loss + Fisheries assets damage (complete and partial)

→ **5.1 Fisheries Production Damage is composed of the:**

1) Pre-disaster value of stored inputs: $\Delta q_{x(stored)j,t} \cdot P_{x(stored)j,t-1}$

2) Pre-disaster value of destroyed capture: $\Delta q_{i(stored)j,t} \cdot P_{i(stored)j,t-1}$

$$PD(FI)_{ij} = \Delta q_{x(stored)j,t} \cdot P_{x(stored)j,t-1} + \Delta q_{i(stored)j,t} \cdot P_{i(stored)j,t-1}$$

→ **5.2 Fisheries Production Loss is composed of the:**

1) Difference between expected and actual value of fisheries capture in disaster year: $area_{ij,t} \cdot p_{ij,t-1} \cdot \Delta y_{ij,t}$

$$PL(FI)_{ij} = area_{ij,t} \cdot p_{ij,t-1} \cdot \Delta y_{ij,t}$$

→ **5.3 Fisheries Assets Damage is composed of the:**

1) Repair / replacement cost of partially / fully destroyed assets at pre-disaster price: $p_{kj,t-1} \cdot \Delta q_{kj,t}$

$$AD(ALL)_{ij} = p_{kj,t-1} \cdot \Delta q_{kj,t}$$

Note 1: All prices used in the below computations are pre-disaster farm gate/producer prices.

Note 2: Notation

i	is output
j	is geographical units affected by the disaster
k	is asset (equipment, machinery, tools, facilities) used to produce an agricultural output
x	is input used for agricultural production
h	is trees (perennial crop trees and forest trees)
t	is the first time unit when post-disaster data are available
$t-1$	is the first time unit when pre-disaster data are available
$y_{i,j,t}$	is the yield of item i in zone j at time t
$P_{x(\text{or } i \text{ or } h),j,t-1}$	is the price of input x (or product i or tree h) in zone j at time t-1
$P_{kj,t}$	is the price (or repair cost) of one unit of asset k in zone j at time t
$q_{i,j}$	is the quantity of item i in zone j
$q_{i \text{ (or } x)(\text{stored})j \; tj}$	is the stored quantity of item i (or input x) in zone j at time t
$q_{kj,t}$	is the number of assets used for item i in zone j at time t
$ha_{ij,t}$	is the number of hectares devoted to item i in zone j at time t
$\Delta ha_{ij,t} = E_{t-1}[ha_{ij,t}]-ha_{ij,t}$	is the unexpected change in the number of hectares where i is produced
w_i	is the average weight (in tons) of item i
$P_{(\text{short run})}$	is the lump sum of expenses used to temporarily sustain production activities after a disaster
a	is the share of the value of dead animals that can be sold
$area_{ij,t}$	is the number of units of area where item i (i.e. type of fish) in zone j at time t is caught/bred

Note 3:

It is important to note that this methodology could easily incorporate a resilience dimension, accounting for the specific context in which it is used. Resilience parameters would indicate a higher reconstruction cost in areas where resilience is lower. This would be of particular relevance in the estimation of damage to assets employed in all sub-sectors.

Resilience parameters can be obtained, for instance, by incorporating reconstruction time and costs through indices such as the Vulnerability and Lack of coping capacity dimensions of the Index for Risk Management (INFORM), or other indices such as the Resilience Index Measurement and Analysis (RIMA), which are open-source methodologies for quantitatively assessing crises and disasters risk. The higher the risk, as defined by such indices, the higher would be, ceteris paribus, the cost attached to the disaster in a specific area, given similar hazard intensity.

It should also be noted that certain aspects of resilience are already endogenously incorporated into the methodology through the variability in yields.

Minimum Data Requirements for the computation of damage and loss in agriculture

The computation method presented provides a large degree of flexibility regarding data requirements: it can function with variable degrees of data availability. Below are the minimum data requirements necessary for a functional damage and loss assessment in each subsector.

Data to be collected for each disaster as a minimum requirement for the application of FAO's Damage and Loss Assessment Methodology:

1. DL-C: Number of hectares of crops damaged/destroyed by disasters and associated damaged/destroyed machinery and facilities, disaggregated by:
→ types of cultivated crops in the affected areas;
→ types of damaged machinery and facilities;
→ number of fully and/or partially damaged hectares, infrastructure and facilities.

2. DL-L: Number of livestock killed or affected by disasters and associated damaged/destroyed machinery and facilities, disaggregated by:
→ types of livestock;
→ type of damaged infrastructure and facilities;
→ number of dead or affected livestock;
→ number of fully or partially damaged machinery and facilities.

3. DL-FO: Number of hectares of forests damaged/destroyed by disasters, and associated damaged/destroyed machinery and facilities, disaggregated by:
→ main type of forest;
→ types of damaged infrastructure and facilities;
→ number of fully and/or partially damaged hectares, machinery and facilities.

4. DL-AQ: Aquaculture production area affected and associated damaged/destroyed machinery and facilities, disaggregated by:
→ types of aquaculture activities in affected areas;
→ types of damaged infrastructure and facilities;
→ number of fully or partially damaged machinery and facilities.

5. DL-FI: Fisheries production area affected, and associated damaged/destroyed machinery and facilities, disaggregated by:
→ types of fishing activities in the affected areas;
→ types of damaged infrastructure and facilities, primarily vessels;
→ number of fully and/or partially damaged machinery and facilities, primarily vessels.

Error analysis and margin of error

The proposed computation methods are based on a set of assumptions and exogenous knowledge-based parameters, which are oriented towards a conservative approach. Results however might be biased for a variety of reasons.

First, the lack of data (both pre- and post-disaster) and the impossibility of relaxing the assumptions implies the utilization of estimation procedures. Second, errors may occur due to distortions and simultaneous causes of changes in agricultural outputs, other than the natural hazard. Third, lack of sensitivity in the measurement may be a significant source of bias.

Finally, the knowledge-based features of the computation method may modify the output depending on the source of knowledge. In order to mirror this variability in the statistics provided for damage and loss values from natural disasters, a two-step error analysis could be proposed. The first step considers the variability in the definition of the exogenous parameters; the second may be used to test the robustness of the average disaster impact in agriculture across levels of the climatic stressors.

If necessary, the following proposed error interval procedures may be applied in order to represent at least part of the variability in the outcome measurements.

1. Min-Max interval. The computation method presents a set of exogenous (estimated) data in each sub-component, distinctly for damage and for loss.

→ An average, a minimum and a maximum value is defined for each of the data estimations. All three values are primarily based on the existing literature and available expert judgment.

→ The outcome values for damage and loss are calculated three times for each sub-component, using the average values of the exogenous data, the values that minimize the outcome, and the values that maximize the outcome.

2. Confidence interval per level of geophysical stressor. In order to identify the magnitude of a disaster, climatic and geophysical stressor information should be collected at the most cost-efficient available level of granularity.

→ Categories of intensity of the stressors should be defined. For instance, in the case of hurricanes, wind speed is a strong determinant of the magnitude of the natural hazard, and five categories can be identified.

→ For each cluster (i.e. category of stressor's intensity), the mean of damage and of loss in zones j falling under that precise cluster should be calculated.

→ Each mean should be provided with a 90 percent or 95 percent confidence interval. Hypothesis test of difference between means should then be calculated. The T test tests overall internal validity.

A detailed set of methodological guidance notes and a practitioner's tool kit, aimed to assist with the practical application of the methodology in the assessment of post-disaster damage and loss in the agriculture sector, is currently under development by the FAO and will be made available soon.

Glossary

Agricultural assets: The volume of stored inputs and production (seeds, fertilizer, feed, stored crops and livestock produce, harvested fish, stored wood, etc.) as well as machinery and equipment used in crop and livestock farming, forestry, aquaculture and fisheries (includes, but is not limited to: tractors, balers, combine harvesters, threshers, fertilizer distributors, ploughs, root or tuber harvesting machines, seeders, soil machinery, irrigation facilities, tillage implements, track-laying tractors, milking machines, dairy machines, wheeled special machines, portable chain-saws, fishing vessels, fishing gear, aquaculture feeders, pumps and aerators, aquaculture support vessels).

Agricultural production loss: Declines in the volume of crop, livestock (and also forestry, aquaculture and fisheries) production resulting from a disaster, as compared to pre-disaster expectations.

Baseline (or baseline assessment): Pre-disaster information, which includes national or sub-national, data relevant to the disaster-affected areas, incl uding indicators such as yields, production volume, prices, malnutrition and food insecurity, income levels. Comparisons based on baseline data are critical to determining the overall impact of the disaster.

Biological disasters: Are of organic origin or conveyed by biological vectors, including pathogenic microorganisms, toxins and bioactive substances. Examples are bacteria, viruses or parasites, as well as venomous wildlife and insects, poisonous plants and mosquitoes carrying disease-causing agents (UNISDR 2017).

Climatological disasters: A disaster caused by long-lived, meso- to macro-scale atmospheric processes ranging from intra-seasonal to multi-decadal climate variability (EM-DAT CRED 2017).

Conflicts: Situations of civil unrest, regime change, interstate conflicts, civil wars, etc.

Damage: The monetary value of total or partial destruction of physical assets and infrastructure in disaster-affected areas, expressed as replacement and/or repair costs. In the agriculture sector, damage is considered in relation to standing crops, farm machinery, irrigation systems, livestock shelters, fishing vessels, pens and ponds, etc. (EU, UNDG & World Bank 2013, UNISDR 2017, FAO 2017a).

Disaster: A serious disruption of the functioning of a community or a society at any scale due to hazardous events interacting with conditions of exposure, vulnerability and capacity, leading to one or more of the following: human, material, economic and environmental loss and impacts (UNISDR 2017).

Disaster risk reduction (DRR): The policy objective aimed at preventing new and reducing existing disaster risk and managing residual risk, all of which contribute to strengthening resilience and achievement of sustainable development (UNISDR 2017).

Early-warning system: An integrated system of hazard monitoring, forecasting and prediction, disaster risk assessment, communication and preparedness activities systems and processes that enables individuals, communities, governments, businesses and others to take timely action to reduce disaster risks in advance of hazardous events (UNISDR 2017).

Food chain: The series of processes by which food is grown or produced, sold, and eventually consumed.

Food chain crises: Threats to the human food chain, such as transboundary plant, forest, animal, aquatic and zoonotic pests and diseases, food safety events, radiological and nuclear emergencies, dam failures, industrial pollution, oil spills, etc. These have the potential to significantly affect food security, livelihoods, human health, national economies and global markets (FAO 2017).

Food security and nutrition: A situation that exists when all people at all times have physical, social and economic access to sufficient, safe and nutritious food that meets their dietary needs and food preferences for an active and healthy life.

Geophysical disasters: Originate from the Earth's internal processes. Examples are earthquakes, volcanic activity and emissions, and related geophysical processes such as mass movements, landslides, rockslides, surface collapses and debris or mud flows. Hydro- and meteorological factors are important contributors to some of these processes. Tsunamis are difficult to categorize: although they are triggered by undersea earthquakes and other geological events, they essentially become an oceanic process that is manifested as a coastal water-related hazard (UNISDR 2017).

Hazard: a process or phenomenon that may cause loss of life, injury or other health impacts, property damage, social and economic disruption or environmental degradation (UNISDR 2017). Hazards may be natural, anthropogenic or socio-natural in origin; this report refers to hazards of natural origin only. Natural hazards are predominantly associated with natural processes and phenomena.

Hydrological disasters: A disaster caused by the occurrence, movement, and distribution of surface and subsurface freshwater and saltwater (EM-DAT CRED, 2017).

Loss: The change in economic flows occurring as a result of a disaster. In agriculture, loss may include declines in crop production, decline in income from livestock products, increased input prices, reduced overall agricultural revenues and higher operational costs and increased unexpected expenditure to meet immediate needs in the aftermath of a disaster (EU, UNDG & World Bank 2013, UNISDR 2017, FAO 2017a).

Meteorological disasters: Events caused by short-lived/small-to mesoscale atmospheric processes (in the spectrum from minutes to days) (EM-DAT CRED 2017).

Migration: The movement of a person or a group of persons, either across an international border or within a state. It is a population movement, encompassing any kind of movement of people, whatever its length, composition and causes. It includes migration of refugees, displaced persons, economic migrants, and persons moving for other purposes, including family reunification (IOM 2017).

Mitigation: The lessening or minimizing of the adverse impacts of a hazardous event (UNISDR 2017).

Preparedness: The knowledge and capacities developed by governments, response and recovery organizations, communities and individuals to effectively anticipate, respond to and recover from the impacts of a likely, imminent or current disaster (UNISDR 2017).

Protracted crisis: Environment in which a significant proportion of the population is acutely vulnerable to death, disease and disruption of livelihoods over a prolonged period of time. The governance of such an environment is usually very weak, with the state having a limited capacity to respond to, or mitigate, threats to the population, or to provide adequate levels of protection (FAO 2010).

Reconstruction: The medium- and long-term rebuilding and sustainable restoration of resilient critical infrastructures, services, housing, facilities and livelihoods required for the full functioning of a community or a society affected by a disaster, aligning with the principles of sustainable development and "building back better," to avoid or reduce future disaster risk (UNISDR 2017).

Recovery: The restoring or improving of livelihoods and health, as well as economic, physical, social, cultural and environmental assets, systems and activities, of a disaster-affected community or society, aligning with the principles of sustainable development and "build back better," to avoid or reduce future disaster risk (UNISDR 2017).

Rehabilitation: The restoration of basic services and facilities for the functioning of a community or a society affected by a disaster (UNISDR 2017).

Resettlement: The relocation and integration of people (refugees, internally displaced persons, etc.) into another geographical area and environment, usually in a third country. In the refugee context, the transfer of refugees from the country in which they have sought refuge to another State that has agreed to admit them (IOM 2017).

Resilience: The ability of a system, community or society exposed to hazards to resist, absorb, accommodate, adapt to, transform and recover from the effects of a hazard in a timely and efficient manner, including through the preservation and restoration of its essential basic structures and functions through risk management (UNISDR 2017).

Response: Actions taken directly before, during or immediately after a disaster in order to save lives, reduce health impacts, ensure public safety and meet the basic subsistence needs of the people affected (UNISDR 2017).

Risk: The potential loss of life, injury, or destroyed or damaged assets which could occur to a system, society or a community in a specific period of time, determined probabilistically as a function of hazard, exposure, vulnerability and capacity. The definition of disaster risk reflects the concept of hazardous events and disasters as the outcome of continuously present conditions of risk (UNISDR, 2017).

Bibliography

ACAPS. 2013. Impact of the conflict on Syrian economy and livelihoods. Syria Needs Analysis Project – July 2013. Geneva, ACAPS. 23 pp. (also available at https://www.acaps.org/sites/acaps/files/products/files/23_impact_of_the_conflict_on_syrian_economy_and_livelihoods_july_2013.pdf).

Adhikari, C., Adhikary, B., Rajbhandari, N.P., Hooper, M., Upreti, H.K., Gyawali, B.K., Rajbhandari, N.K. & Hobbs, P.R. 1999. Wheat and Rice in the Mid-Hills of Nepal: A Benchmark Report on Farm Resources and Production Practices in Kavre District. Nepal Agricultural Research Council (NARC), Agronomy Division, Khumaltar, Nepal, International Maize and Wheat Improvement Center (CIMMYT). 36 pp. (also available at http://repository cimmyt.org/xmlui/bitstream/handle/10883/3937/68337.pdf?sequence=1&isAllowed=y).

Aimin, H. 2010. Uncertainty, Risk Aversion and Risk Management in Agriculture. Agriculture and Agricultural Science Procedia, 1: 152-156. (also available at https://ac.els-cdn.com/S2210784310000197/1-s2.0-S2210784310000197-main.pdf?_tid=47e9fa56-c2d6-11e7-aeae-00000aab0f27&acdnat=1509961304_5210f5dc12326690e9bfbbf96ef9ad89).

Allen C.D., Macalady, A.K., Chenchouni, H., Bachelet, D., McDowell, N., Vennetier, M., Kitzberger, T., Rigling, A., Breshears, D.D., Hogg, E.H., Gonzales, P., Fensham, R., Zhang, Z., Castro, J., Demidova, N., Lim, J.H., Allard, G., Running, S.W., Semerci, A. & Cobb N. 2010. A global overview of drought and heat-induced tree mortality reveals emerging climate change risks to forests. Forest Ecology and Management 259 (4): 660-684. (also available at http://www.sciencedirect.com/science/article/pii/S037811270900615X).

Anyamba, A., Linthicum, K.J., Small, J.L., Collins, K.M., Tucker, C.J., Pak, E.W., Britch, S.C., Eastman, J.R., Pinzon, J.E. & Russel, K.L. 2012. Climate Teleconnections and Recent Patterns of Human and Animal Disease Outbreaks. PLOS Neglected Tropical Diseases 6(1): 1-14. (also available at http://journals.plos.org/plosntds/article?id=10.1371/journal.pntd.0001465).

Baniya, B.K., Tiwari, R.K., Chaudhary, P., Shrestha, S.K. & Tiwari, P.R. 2005. Planting Materials Seed Systems of Finger Millet, Rice and Taro in Jumla, Kaski and Bara Districts of Nepal. Nepal Agriculture Research. Journal, 6(2005): 39-48. (also available at http://www.nepjol.info/index.php/NARJ/article/view/3343).

Bjorheden, R. 2007. Possible effects of the hurricane Gudrun on the regional Swedish forest energy supply. Biomass and Energy 31(9): 617-622. (also available at http://www.sciencedirect.com/science/article/pii/S0961953407000979).

Bloesch, U. 2016. Impact de l'ouragan Matthew sur le secteur forestier. Possibilités de valorisation de produits forestiers et nécessité de réhabilitation des écosystèmes après l'ouragan Matthew en Haïti (2-22/12/2016). Rome, FAO. 32 pp. (also available at http://www.adansonia-consulting.ch/document/Rapport%20final%20Bloesch.%20Impact%20de%20l%27ouragan%20Matthew%20sur%20le%20secteur%20forestier.pdf).

Checchi, F. & Robinson, W.C. 2013. Mortality among populations of southern and central Somalia affected by severe food insecurity and famine during 2010-2012. Washington, DC and Rome, FSNAU & FEWSNET. 87 pp. (also available at http://www.fsnau.org/in-focus/study-report-mortality-among-populations-southern-and-central-somalia-affected-severe-food-).

Cole, S., Gine, X., Tobacman, J., Topalova, P., Townsend, R., Vickery, J. 2013. Barriers to Household Risk Management: Evidence from India. American Economic Journal: Applied Economics, 5: 104–135. (also available at https://openknowledge.worldbank.org/bitstream/handle/10986/17926/AEJApp_2011_0326_Published.pdf;sequence=1).

COP19. 2013. Warsaw International Mechanism for Loss and Damage associated with Climate Change Impacts. 4 pp. (also available at http://unfccc.int/files/meetings/warsaw_nov_2013/in-session/application/pdf/fccc.cp.2013.l.15.pdf).

COP21. 2015. The Paris Agreement. 27pp. (also available at http://unfccc.int/files/essential_background/convention/application/pdf/english_paris_agreement.pdf).

Cosic, D., Bajracharya, R.D., Dahal, S., Rana, S. & Shamsher, J.B. 2016. Nepal development update: remittances at risk. Washington, DC, World Bank. 36 pp. (also available at http://documents.worldbank.org/curated/en/564551468198011442/pdf/106393-WP-PUBLIC-ADD-SERIES-Nepal-Development-Update-2016.pdf).

Dar, O., McIntyre, S, Hogarth, S. & Heymann, D. 2013. Rift Valley Fever and a New Paradigm of Research and Development for Zoonotic Disease Control. Emerging Infectious Diseases 19(2): 189–193. (also available at https://wwwnc.cdc.gov/eid/article/19/2/12-0941_article).

De Haen, H., Hemrich, G. 2007. The Economics of Natural Disasters: Implications and Challenges for Food Security. Agricultural Economics, 37(1): 31-45. (also available at http://onlinelibrary.wiley.com/doi/10.1111/j.1574-0862.2007.00233.x/full).

Development Initiatives. 2016a. Forced displacement poverty and financing: Seven facts you need to know. Bristol, Development Initiatives. 9 pp. (also available at http://devinit.org/wp-content/uploads/2016/09/Forced_Displacement_Poverty_and_Financing_DI_Sept_2016.pdf).

Development Initiatives. 2016b. Global Humanitarian Assistance Report 2016.Bristol, Development Initiatives. 101 pp. (also available at http://devinit.org/wp-content/uploads/2016/06/Global-Humanitarian-Assistance-Report-2016.pdf).

ECLAC. 2014. Handbook for Disaster Assessment, 3rd edition. Santiago, Chile, UN. 300 pp. (also available at http://repositorio.cepal.org/bitstream/handle/11362/36823/1/S2013817_en.pdf).

El Laithy, H. & Abu-Ismail, K. 2005. Poverty in Syria: 1996-2004. Diagnosis and Pro-Poor Policy Considerations. UNDP, Syria. 146 pp. (also available at http://dspace.africaportal.org/jspui/bitstream/123456789/20376/1/Poverty%20In%20Syria%201996%202004%20Diagnosis%20And%20Pro%20Poor%20Policy%20Considerations.pdf?1).

ESCWA. 2016. Syria at War Five Years On. 36 pp. (also available at https://www.unescwa.org/sites/www.unescwa.org/files/publications/files/syria-war-five-years.pdf).

ESCWA. Jan. 2017. Strategic Policy Alternatives Framework (SPAF) Syria Post-Conflict. 44 pp. (also available at http://www.rawabet.org/sites/default/files/03-spaf_synopsis-_3.pdf).

EU, UNDG & World Bank. 2013. Post-Disaster Needs Assessments. Volume A. Guidelines. 126 pp. (also available at http://www.undp.org/content/dam/undp/library/Environment%20and%20Energy/Climate%20Strategies/PDNA%20Volume%20A%20FINAL%2012th%20Review_March%202015.pdf).

EU, UNDG & World Bank. 2014. Agriculture, Livestock, Fisheries and Forestry. PDNA Guidelines Volume B. 49 pp. (also available at http://documents.worldbank.org/curated/en/341131493098412619/pdf/114518-WP-PUBLIC-ADD-SERIES-pdna-guidelines-vol-b-agriculture-livestock-fisheries-forestry-0.pdf).

FAO & Nepal Food Security Cluster. 2015. Nepal Earthquake, Agricultural Livelihood Impact Appraisal in Six Most Affected Districts. 54 pp. (also available at http://fscluster.org/sites/default/files/documents/Nepal%20ALIA%20-%20Agricultural%20Livelihoods%20Impact%20Appraisal%20-%20June%2006.pdf).

FAO & WFP. 2011. Joint Rapid Food Security Needs Assessment (JRFSNA) Syrian Arab Republic. Report. June 2012. Rome. 27 pp. (also available at http://documents.wfp.org/stellent/groups/public/documents/ena/wfp250023.pdf?_ga=2.252520073.1088213805.1508831705-429038838.1508422336).

FAO, Government of Myanmar, WFP. 2015. Agriculture and Livelihood Flood Impact Assessment in Myanmar. 60 pp. (also available at http://www.fao.org/fileadmin/user_upload/emergencies/docs/Final_Impact_Assessment_Report_final.pdf).

FAO. 1995. Code of Conduct for Responsible Fisheries. Rome, FAO. 41 pp. (also available at http://www.fao.org/docrep/005/v9878e/v9878e00.HTM).

FAO. 2004. Global Forest Resources Assessment Update 2005 – Terms and Definitions. Working Paper 83/E. Rome, FAO. 34 pp. (also available at http://www.fao.org/forestry/7797-0f7ba44a281b061b9c964d3633d8bf325.pdf).

FAO. 2009. State of Food and Agriculture 2009: Livestock in the Balance. Rome, FAO. 180 pp. (also available at http://www.fao.org/docrep/012/i0680e/i0680e.pdf).

FAO. 2010. State of Food Insecurity in the World: Addressing Food Insecurity in Protracted Crises. Rome, FAO. 62pp. (also available at http://www.fao.org/docrep/013/i1683e/i1683e.pdf).

FAO. 2013a. Guidelines for the Fisheries and Aquaculture Sector on Damage and Needs Assessments in Emergencies. Rome, FAO. 125 pp. (also available at http://www.fao.org/3/a-i3433e.pdf).

FAO. 2013b. Resilient Livelihoods. Disaster Risk Reduction for Food and Nutrition Security. Rome, FAO. 108 pp. (also available at http://www.fao.org/3/a-i3270e.pdf).

FAO. 2015a. Global Forest Resources Assessment 2015. Desk Reference. Rome, FAO. 253 pp. (http://www.fao.org/3/a-i4808e.pdf).

FAO. 2015b. The Impact of Disasters on Agriculture and Food Security. Rome, FAO. 54 pp. (also available at http://www.fao.org/3/a-i5128e.pdf).

FAO. 2015c. The State of Food Insecurity in the World. Meeting the 2015 international hunger targets: taking stock of uneven progress. Rome, FAO. 62 pp. (also available at http://www.fao.org/3/a-i4646e.pdf).

FAO. 2015d. Voluntary Guidelines for Securing Sustainable Small-Scale Fisheries in the Context of Food Security and Poverty Eradication. Rome, FAO. 34 pp. (also available at http://www.fao.org/3/a-i4356e.pdf).

FAO. 2016. 2015–2016 El Niño. Early action and response for agriculture, food security and nutrition. Rome, FAO. 43 pp. (also available at http://www.fao.org/emergencies/resources/documents/resources-detail/it/c/340660/).

FAO. 2016. Global Forest Resources Assessment 2015 – How are the world's forests changing? Second edition. Rome, FAO. 54 pp. (http://www.fao.org/3/a-i4793e.pdf).

FAO. 2017a. Counting the Cost: Agriculture in Syria after Six Years of Crisis. 20 pp. (also available at http://www.fao.org/emergencies/resources/documents/resources-detail/en/c/878213/).

FAO. 2017b. Supporting the energy needs of refugees and host communities: the role of forests [video]. [Cited 1 Dec. 2017]. https://www.youtube.com/watch?v=ysIoC6vO4A4&feature=youtu.be.

FAO. 2017c. The State of Food Security and Nutrition in the World. Building Resilience for Peace and Food Security. Rome, FAO. 132pp. (also available at http://www.fao.org/3/a-I7695e.pdf).

Frenken, K. ed. 2009. Irrigation in the Middle East region in figures. AQUASAT Survey – 2008. Rome, FAO. 423 pp. (also available at http://www.fao.org/3/a-i0936e.pdf).

Garrasi, D. & Allen, R. 2016. Review of the Experiences with Post Conflict Needs Assessments: 2008-15. 81 pp. (also available at https://openknowledge.worldbank.org/bitstream/handle/10986/24029/K8699.pdf?sequence= 2&isAllowed=y).

Garschagen, M., Hagenlocher, M., Kloos, J., Pardoe, J., Lanzendörfer, M., Mucke, P., Radtke, K., Rhyner, J., Walter, B., Welle, T. & Birkmann, J. 2015. World Risk Report 2015. Bündnis Entwicklung Hilft and UNU-EHS. 74 pp. (also available at https://collections.unu.edu/eserv/UNU:3303/WRR_2015_engl_online.pdf).

Ghimire, R., Wen-chi, H. & Shrestha, R.B. 2015. Factors Affecting Adoption of Improved Rice Varieties among Rural Farm Households in Central Nepal. ScienceDirect, Rice Science, 2015, 22(1): 35-43. (also available at http://www.sciencedirect.com/science/article/pii/S1672630815000074).

Government of Bosnia and Herzegovina. 2014. Bosnia and Herzegovina Floods, 2014. Recovery Needs Assessment. 302 pp. (also available at http://www.ilo.org/wcmsp5/groups/public/---ed_emp/documents/publication/wcms_397687.pdf).

Government of Cabo Verde. 2015. Post-Disaster Needs Assessment (PDNA). Fogo Volcanic Eruption 2014-2015. 127 pp. (also available at https://www.un.cv/files/PDNAREPORT_EN.PDF).

Government of Fiji. 2016. Fiji. Post Disaster Needs Assessment. Tropical Cyclone Winston, February 20, 2016. 160 pp. (also available at https://www.gfdrr.org/sites/default/files/publication/Post%20Disaster%20Needs%20Assessments%20CYCLONE%20WINSTON%20Fiji%202016%20%28Online%20Version%29.pdf).

Government of Haiti. 2016. Evaluation des besoins post ouragan Matthew dans le secteur agricole. 50 pp. (also available at http://www.ht.undp.org/content/dam/haiti/docs/Prevention%20des%20crises/rapportsectorielPD-NA/UNDP-HT-Rapport-sectoriel-agriculture-VF-sm.pdf).

Government of Kenya. 2012. Kenya Post Disaster Needs Assessment (PDNA). 2008-2011 Drought.192 pp.

(also available at http://www.gfdrr.org/sites/gfdrr/files/Kenya_PDNA_Final.pdf).

Government of Malawi. 2015. Malawi 2015 Floods Post Disaster Needs Assessment Report. 111 pp. (also available at http://reliefweb.int/sites/reliefweb.int/files/resources/Malawi-2015-Floods-Post-Disaster-Needs-Assessment-Report.pdf).

Government of Marshall Islands. 2017. Post Disaster Needs Assessment of the 2015-2016 Drought. 138 pp. (also available at http://www.ilo.org/wcmsp5/groups/public/---ed_emp/documents/publication/wcms_553635.pdf).

Government of Myanmar. 2015. Myanmar. Post-Disaster Needs Assessment of Floods and Landslides July-September 2015. 302 pp. (also available at http://earlyrecovery.global/sites/default/files/myanmar000postjuly000september2015_1_0.pdf).

Government of Nepal National Planning Commission. 2013. National Sample Census of Agriculture, Nepal, 2011/12. 129 pp. (also available at http://www.fao.org/fileadmin/templates/ess/ess_test_folder/World_Census_Agriculture/Country_info_2010/Reports/Reports_5/NPL_EN_REP_2011-12.pdf).

Government of Nepal National Planning Commission. 2015a. Nepal Earthquake 2015 Post Disaster Needs Assessment, vol. A: Key Findings. 134 pp. (also available at http://www.npc.gov.np/images/category/PDNA_Volume_A.pdf).

Government of Nepal National Planning Commission. 2015b. Nepal Earthquake 2015 Post Disaster Needs Assessment, vol. B: Sector Reports. 348 pp. (also available at http://www.npc.gov.np/images/category/PDNA_volume_BFinalVersion.pdf).

Government of Nepal, OCHA, WFP, FAO, UNDP, Nepal Red Cross Society, Food Security Cluster, Early Recover and Protection Cluster, inter-cluster gender working group, & REACH. 2015. Nepal Earthquake Response. Joint Assessment of Food Security, Livelihoods and Early Recovery. 75 pp. (also available at http://fscluster.org/sites/default/files/documents/nepal_report_joint_assessment_of_food_security_livelihoods_and_early_recovery_2015_13-jan-16.pdf).

Government of Seychelles. 2016. Seychelles Post Disaster Needs Assessment. Tropical Cyclone Fantala. 72 pp. (also available at https://www.gfdrr.org/sites/default/files/Seychelles-Fantala-PDNA.pdf).

Government of Sri Lanka. 2016. Sri Lanka. Post-Disaster Needs Assessment. Floods and Landslides-May 2016. 300 pp. (also available at http://reliefweb.int/sites/reliefweb.int/files/resources/pda-2016-srilanka.pdf).

Government of the Philippines, NDRRMC. 2013. Final Report re: Effects of Typhoon "Yolanda" (Haiyan). Quezon City, Philippines, NDRRMC. 148 pp. (also available at http://ndrrmc.gov.ph/attachments/article/1329/FINAL_REPORT_re_Effects_of_Typhoon_YOLANDA_(HAIYAN)_06-09NOV2013.pdf).

Government of Vanuatu. 2015. Post-Disaster Needs Assessment. Tropical Cyclone Pam, March 2015. 172 pp. (also available at http://www.ilo.org/wcmsp5/groups/public/---ed_emp/documents/publication/wcms_397678.pdf).

Hales, S., Edwards, S.J. & Kovats, R.S. Impacts on health of climate extremes. In McMichael, A.J., Campbell-Lendrum, D.H., Corvalán, C.F., Ebi, K.L., Githeko, A.K., Scheraga, J.D. & Woodward, A., eds. Climate Change and Human Health. Risks and Responses, pp 79-102. Genva, WHO, 2003. (also available at http://www.who.int/globalchange/publications/climchange.pdf).

Hallegatte, S., Bangalore, M. & Nkoka, F. 2015. Recent Floods in Malawi Hit the Poorest Areas: What This Implies. Retrieved from Voices-Perspectives on Development. [Cited 1 Dec. 2017]. https://blogs.worldbank.org/voices/recent-floods-malawi-hit-poorest-areas-what-implies.

Hallegatte, S., Bangalore, M., Bonzanigo, L., Fay, M., Kane, T., Narloch, U., Rozenberg, J., Treguer, D. & Vogt-Schilb, A. 2015. Poverty and climate change: natural disasters, agricultural impacts and health shocks. In Barrett, S., Carraro, C. & De Melo, J., eds. 2015. Towards a workable and effective climate regime, pp. 369-389. London, Centre for Economic Policy Research. (also available at http://voxeu.org/sites/default/files/image/FromMay2014/Climate%20change%20book%20for%20web.pdf).

Hallegatte, S., Vogt-Schilb, A., Bangalore, M. & Rozenberg, J. 2017. Unbreakable: Building the Resilience of the Poor in the Face of Natural Disasters. Climate Change and Development Series. Washington, DC, World Bank. 27 pp. (also available at https://openknowledge.worldbank.org/bitstream/handle/10986/25335/211003ovEN.pdf).

Hoddinott, J. & Kinsey, W. 2001. Child growth in the time of drought. Oxford Bulletin of Economics and Statistics, 63(4): 409–436. (also available at https://are.berkeley.edu/courses/ARE251/2004/papers/Hoddinott_Kinsey.pdf).

Hossain, M.Z., Islam, M.T., Sakai, T. & Ishida, M. Impact of tropical cyclones on rural infrastructures in Bangladesh. Agricultural Engineering International: CIGR Journal, 10(2): 1-13. (also available at http://www.cigrjournal.org/index.php/Ejounral/article/view/1036).

IFAD. 2007. Syrian Arab Republic: Thematic study on land reclamation through de-rocking. Rome, IFAD. 61 pp. (https://www.ifad.org/documents/10180/30369bdd-3b95-4760-acd1-39d190584e4c).

IFAD. 2012. Agricultural value chain finance strategy and design. Technical Note. Rome, IFAD. 50 pp. (also available at https://www.ifad.org/documents/10180/8d74b792-33e4-4f57-96b4-e348af035a3c).

IOM. 2017. Key migration terms. [Cited 1 Dec. 2017.] https://www.iom.int/key-migration-terms.

Israel, D.C., Briones, R.M. 2013. Impacts of Natural Disasters on Agriculture, Food Security, and Natural Resources and Environment in the Philippines. ERIA Discussion Paper Series, ERIA-DP-2013-15. 54 pp. (also available at http://www.eria.org/ERIA-DP-2013-15.pdf).

Lagmay, A.M., Agaton, R.P., Bahala, M.A.C. & Tablazon, J.P. 2014. Devastating Storm Surges of Typhoon Yolanda. Project NOAH Open-File Reports, vol. 3, n. 6: 45-56. (also available at http://d2lq12osnvd5mn.cloudfront.net/SS_yolanda.pdf).

Lamichhane, J., Timsina, K.P., RanaBhat, D.B. & Adhikari, S. 2015. Technology adoption analysis of improved maize technology in western hills of Nepal. Journal of Maize Research and Development (2015) 1(1): 146-152. (also available at https://www.nepjol.info/index.php/JMRD/article/view/14252).

Lindenmayer, D.B., Burton, P.J. & Franklin, J.F. 2008. Salvage Logging and its Ecological Consequences. Washington, DC, Island Press. 246 pp.

McAdoo, B., Ah-Leong, J., Bell, L., Ifopo, P., Ward, J., Lovell, E. & Skelton, P. 2011. Coral reefs as buffers during the 2009 South Pacific tsunami, Upolu Island, Samoa. Earth-Science Reviews 107 (July 2011): 147–155. (also available at http://www.sciencedirect.com/science/article/pii/S0012825210001601).

MEA. 2005. Ecosystems and human well-being – current state and trends. Washington, DC, Covelo, London, Island Press. 964 pp. (also available at https://www.millenniumassessment.org/en/Condition.html).

MOAD, FAO & WFP. 2015. Crop Situation Update. A Joint assessment of 2015/16 summer crops and outlook of 2016 winter crops. 21 pp. (also available at http://neksap.org.np/uploaded/resources/Publications-and-Research/Crop%20Situation%20Update/Crop_Situation_Update_summer_2015-16.pdf).

Munyua,P., Murithi, R.M., Wainwright, S., Githinji, J., Hightower, A., Mutonga, D., Macharia, J., Ithondeka, P.M., Musaa, J., Breiman, R.F., Bloland, P., Njenga, M.K. 2010. Rift Valley fever outbreak in livestock in Kenya, 2006-2007. The American Journal of Tropical Medicine and Hygiene, 83: 58–64. (also available at https://www.researchgate.net/profile/Peninah_Munyua2/publication/46054534_Rift_Valley_Fever_Outbreak_in_Livestock_in_Kenya_2006-2007/links/0c960528a1641d5638000000/Rift-Valley-Fever-Outbreak-in-Livestock-in-Kenya-2006-2007.pdf).

N'gang'a, C.M., Bukachi, S.A., Bett, B.K. 2016. Lay perceptions of risk factors for Rift Valley fever in a pastoral community in northeastern Kenya. BMC Public Health 16(32): 1-10. (also available at https://bmcpublichealth.biomedcentral.com/articles/10.1186/s12889-016-2707-8).

Nanyingi, M.O., Munyua, P., Kiama, S.G., Muchemi, G.M., Thumbi, S.M., Bitek, A.O., Bett, B., Muriithi, R.M. & Njenga, M.K. 2015. A systematic review of Rift Valley Fever epidemiology 1931-2014. Infection Ecology & Epidemiology 5: 1-13. (also available at http://dx.doi.org/10.3402/iee.v5.28024).

OCHA. 2011. Eastern Africa Drought Humanitarian Report No. 3, 10 June 2011. 11pp. (also available at https://reliefweb.int/sites/reliefweb.int/files/resources/OCHA%20Eastern%20Africa%20Humanitarian%20Report%20No.%203%20-%20Drought%20May%202011%20FINAL.pdf).

OCHA. 2013. Pacific: Tropical Cyclone Evan. Situation Report No. 9 (as of 11 Jan 2012). 4 pp. (also available at https://reliefweb.int/sites/reliefweb.int/files/resources/TCEvan_SitRep9_11Jan.pdf).

OCHA. 2017. Coordination Tools: Needs Assessments. [Cited 1 Dec. 2017]. https://www.unocha.org/legacy/what-we-do/coordination-tools/needs-assessment.

OECD & World Bank. 2016. Climate and Disaster Resilience Financing in Small Island Developing States. 96 pp. (also available at http://www.oecd-ilibrary.org/development/climate-and-disaster-resilience-financing-in-small-island-developing-states_9789264266919-en).

Orinde, A.B., Kimani, T., Schelling, E., Omolo, J., Kikuvi, G.M. & Njenga, K.M. 2012. Estimation of the Rift Valley Fever burden of disease in the 2006/2007 outbreak in Kenya. Paper presented at the 61st Conference of the American Society of Tropical Medicine and Hygiene, Atlanta, 11-15 Nov. 2012. Nairobi: ILRI. (also available at https://cgspace.cgiar.org/handle/10568/27755).

Oxfam. 2012. Food Crisis in the Horn of Africa, Progress Report July 2011–July 2012. Oxford, Oxfam. 32 pp. (also available at https://www.oxfam.org/sites/www.oxfam.org/files/er-horn-of-africa-2011-2012-progress-report-050712-en.pdf).

Pandey S., Bhandary, H. & Hardy, B. 2007. Economic costs of drought and rice farmers' coping mechanisms. Manila, IRRI. 211 pp. (also available at http://books.irri.org/9789712202124_content.pdf).

Peyre, M., Chevalier, V., Abdo-Salem, S., Velthuis, A., Antoine-Moussiaux, N., Thiry E, Roger, F. 2015. A Systematic Scoping Study of the Socio-Economic Impact of Rift Valley Fever: Research Gaps and Needs. Zoonoses and Public Health 62(5): 309–325. (also available at https://bmcpublichealth.biomedcentral.com/track/pdf/10.1186/s12889-016-2707-8?site=bmcpublichealth.biomedcentral.com).

Purse, B.V., Brown, H.E., Harrup, L., Mertens, P.P.C., Rogers, D.J. 2008. Invasion of bluetongue and other orbivirus infections into Europe: the role of biological and climatic processes. Revue Scientifique et Technique (International Office of Epizootics)27(2): 427–442. (also available at https://www.researchgate.net/publication/23285584_Invasion_of_bluetongue_and_other_orbivirus_infections_into_Europe_The_role_of_biological_and_climatic_processes).

Rhoades, R. 1985. Traditional potato production and farmers' selection of varieties in Eastern Nepal. Potatoes in Food Systems Research Series, Report No. 2. 46 pp. (also available at http://pdf.usaid.gov/pdf_docs/PNAAT487.pdf).

Rich, K.M. & Wanyoike, F. 2010. An Assessment of the Regional and National Socio-Economic Impacts of the 2007 Rift Valley Fever Outbreak in Kenya. The American Journal of Tropical Medicine and Hygiene 83: 52–57. (also available at https://www.ncbi.nlm.nih.gov/pmc/articles/PMC2913501/).

Rosegrant, M.W., Ringler, C. & Zhu, T. 2009. Water for Agriculture: Maintaining Food Security under Growing Scarcity. Annual Review of Environment and Resources, 34: 205-222. (also available at http://www.annualreviews.org/doi/pdf/10.1146/annurev.environ.030308.090351).

Shrestha, A.B., Bajracharya, S.R. & Kargel, J.S. 2016. The Impact of Nepal's 2015 Gorkha Earthquake-Induced Geohazards. ICIMOD Research Report 2016/1. ICIMOD, Kathmandu, Nepal. 48 pp. (also available at http://lib.icimod.org/record/31937/files/icimod-2015-Earthquake-InducedGeohazards.pdf).

Takagi, H. & Esteban, M. 2016. Statistics of tropical cyclone landfalls in the Philippines: unusual characteristics of 2013 Typhoon Haiyan. Journal of the International Society for the Prevention and Mitigation of Natural Hazards, vol. 80, issue 1: 211-222. (also available at https://www.researchgate.net/publication/282150128_Statistics_of_tropical_cyclone_landfalls_in_the_Philippines_unusual_characteristics_of_2013_Typhoon_Haiyan).

Taubenberger, J.K., Hultin, J.V. & Morens, D.M. 2007. Discovery and characterization of the 1918 pandemic influenza virus in historical context. Antiviral Therapy 12: 581—591. (also available at https://www.ncbi.nlm.nih.gov/pmc/articles/PMC2391305/).

Thorn, S., Bässler, C., Brandl, R., Burton, P.J., Cahall, R., Campbell, J.L., Castro, J., Choi, C.-Y, Cobb, T., Donato, D.C., Durska, E., Fontaine, J.B., Gauthier, S., Hebert, C., Hothorn, T., Hutto, R.L., Lee, E.J., Leverkus, A.B., Lindenmayer, D.B, Obrist, M.K., Rost, J., Seibold, S., Seidl, R., Thom, D., Waldron, K., Wermelinger, B., Winter, M.-B., Zmihorski, M. & Müller, J. 2017. Impacts of salvage logging on biodiversity – a meta analysis. Journal of Applied Ecology. DOI: 10.1111/1365-2664.1294. 11 pp. (also available at https://www.researchgate.net/publication/317045815_Impacts_of_salvage_logging_on_biodiversity_A_meta-analysis).

Turner, W.C., Imologhome, P., Havarua, Z., Kaaya, G.P., Mfune, J.K.E., Mpofu, I.D.T. & Getz, W.M. 2013. Soil ingestion, nutrition and the seasonality of anthrax in herbivores of Etosha National Park. Ecosphere 4(1): 1-19. (also available at http://onlinelibrary.wiley.com/doi/10.1890/ES12-00245.1/abstract).

UN. 2015. Transforming Our World: the 2030 Agenda for Sustainable Development. 41 pp. (also available at https://sustainabledevelopment.un.org/content/documents/21252030%20Agenda%20for%20Sustainable%20Development%20web.pdf).

UN. 2017. Report of the Secretary-General: E. Effective coordination of humanitarian assistance efforts. [Cited 1 Dec. 2017]. https://www.un.org/sg/en/content/sg/report-secretary-general-e-effective-coordination-humanitarian-assistance-efforts-0.

UNHCR. 2016. Global Trends. Forced Displacement in 2016. Geneva, UNHCR. 72 pp. (also available at http://www.unhcr.org/5943e8a34.pdf).

UNISDR. 2015a. Making Development Sustainable: The Future of Disaster Risk Management. Global Assessment Report on Disaster Risk Reduction. Geneva, UNISDR. 316 pp. (also available at http://www.preventionweb.net/english/hyogo/gar/2015/en/gar-pdf/GAR2015_EN.pdf).

UNISDR. 2015b. The Sendai Framework for Disaster Risk Reduction 2015-2030. 37 pp. (also available at http://www.unisdr.org/files/43291_sendaiframeworkfordrren.pdf).

UNISDR. 2017. Terminology on Damage Risk Reduction. [Cited 1 Dec. 2017]. https://www.unisdr.org/we/inform/terminology.

Van Lierop, P., Lindquist, E., Sathyapala, S. & Franceschini, G. 2015. Global forest area disturbance from fire, insect pests, diseases and severe weather events. Forest Ecology and Management 352(7): 78-88. (also available at http://www.sciencedirect.com/science/article/pii/S0378112715003369).

Velthuis, A.G.J., Saatkamp, H.W., Mourits, M.C.M., de Koeijer, A.A., Elbers, A.R.W. 2010. Financial consequences of the Dutch bluetongue serotype 8 epidemics of 2006 and 2007. Preventive Veterinary Medicine 93(4): 294–304. (also available at http://www.sciencedirect.com/science/article/pii/S016758770900350X).

WHO & FAO. 2005. Food Safety in Natural Disasters. INFOSAN Information Note No. XXX/2005. 4 pp. (also available at http://www.searo.who.int/entity/emergencies/documents/food_safety_in_disasters.pdf?ua=1).

Wilson, A.J. & Mellor, P.S. 2009. Bluetongue in Europe: past, present and future. Philosophical Transactions of the Royal Society B 364(1530): 2669–2681. (also available at http://rstb.royalsocietypublishing.org/content/364/1530/2669).

World Bank. 2015. Another Nargis Strikes Everyday: Post-Nargis Social Impacts Monitoring Five Years On. 45 pp. Washington, DC, World Bank. (also available at https://www.gfdrr.org/sites/gfdrr/files/publication/Another-Nargis-Strikes-Every-Day.pdf).

World Bank. 2016a. Nepal Development Update, January 2016: Development Amidst Disturbances. 24 pp. Washington, DC, World Bank. (also available at https://openknowledge.worldbank.org/bitstream/handle/10986/23776/Nepal000Develo00amidstodisturbances.pdf?sequence=1&isAllowed=y).

World Bank. 2016b. Recovery and Peacebuilding Assessments (RPBA): FAQs. [Cited 1 Dec. 2017]. http://www.worldbank.org/en/topic/fragilityconflictviolence/brief/recovery-peacebuilding-assessments-faqs.

World Bank. 2016c. The cost of fire: an economic analysis of Indonesia's 2015 fire crisis. Indonesia Sustainable Landscape Knowledge Note: 1. World Bank, Jakarta, Indonesia. 12 pp. (also available at http://documents.worldbank.org/curated/en/776101467990969768/The-cost-of-fire-an-economic-analysis-of-Indonesia-s-2015-fire-crisis).

Databases

EM-DAT CRED

http://www.emdat.be/database

FAOSTAT

http://www.fao.org/faostat/en/#home

Government of Ethiopia Central Statistical Agency

http://www.csa.gov.et/survey-report/category/59-agricultural-sample-survey-belg.html

UNDESA, World Population Prospects 2017

https://esa.un.org/unpd/wpp/

The World Bank World Development Indicators

https://data.worldbank.org/data-catalog/world-development-indicators